Table of Contents

EVERY SUNDAY NIGHT ON 1170 A.M. KFAQ RADIO OUT OF TULSA, OK. I TALK FOR 2 HOURS TO THE BEST FOLKS I KNOW. TO MY LISTENERS, WHO ARE TUNED IN FAITHFULLY EACH WEEK. TO MY CLIENTS AND THEIR HORSES. YOU HAVE ENRICHED MY LIFE BEYOND IMAGINATION. NOT ONLY ARE YOU RESPONSIBLE FOR MAKING ME A BETTER HORSEMAN. YOU'VE MADE ME A BETTER MAN. I CANNOT SEE EVER ACHIEVING A GREATER HONOR AND PRIVILEGE IN MY LIFE. THAN TALKING TO YOU, AND WORKING WITH YOUR HORSES. I WILL ALWAYS AND FOREVER BE IN YOUR DEBT....

Box Stall Diaries: "Stall 13"

This edition published by
"BJ Rickard"

This book designed and produced by

BJ Rickard
P.O. Box 68
Bixby, OK 74008
1-866-STALL13
(782-5513)

Computer Graphics and Typing: Melissa Gentry
Photography and Drawings: Glenda Huckabay
Keith Huckabay

Senior Editor: BJ Rickard

"Total Absolute Control" ☺ BJ Rickard

Introduction

I've read a lot of books in my life about horses. Some were pretty technical, others were entertaining, others were -- -- well maybe you've run into a couple of those books yourself. As for this book, -- I'll just leave that decision to you.

Over the years I have come to look at horses not with a better perspective, but more of a philosophical approach and technique, pertaining to their view and how horses perceive our world. Make no mistake about it there is our world, and then there is the world of the horse.

I'll start by giving you some points of view. These will possibly help you to achieve a more accurate and more fulfilling experience with your horses. Once you have a better idea of how a horse thinks and perceives daily life, then it makes riding and correcting bad habits a whole lot easier.

After this, I will cover what I have found to be the more prominent problems and situations for you and your horses: health issues and just plain old handy tips and information for your consideration.

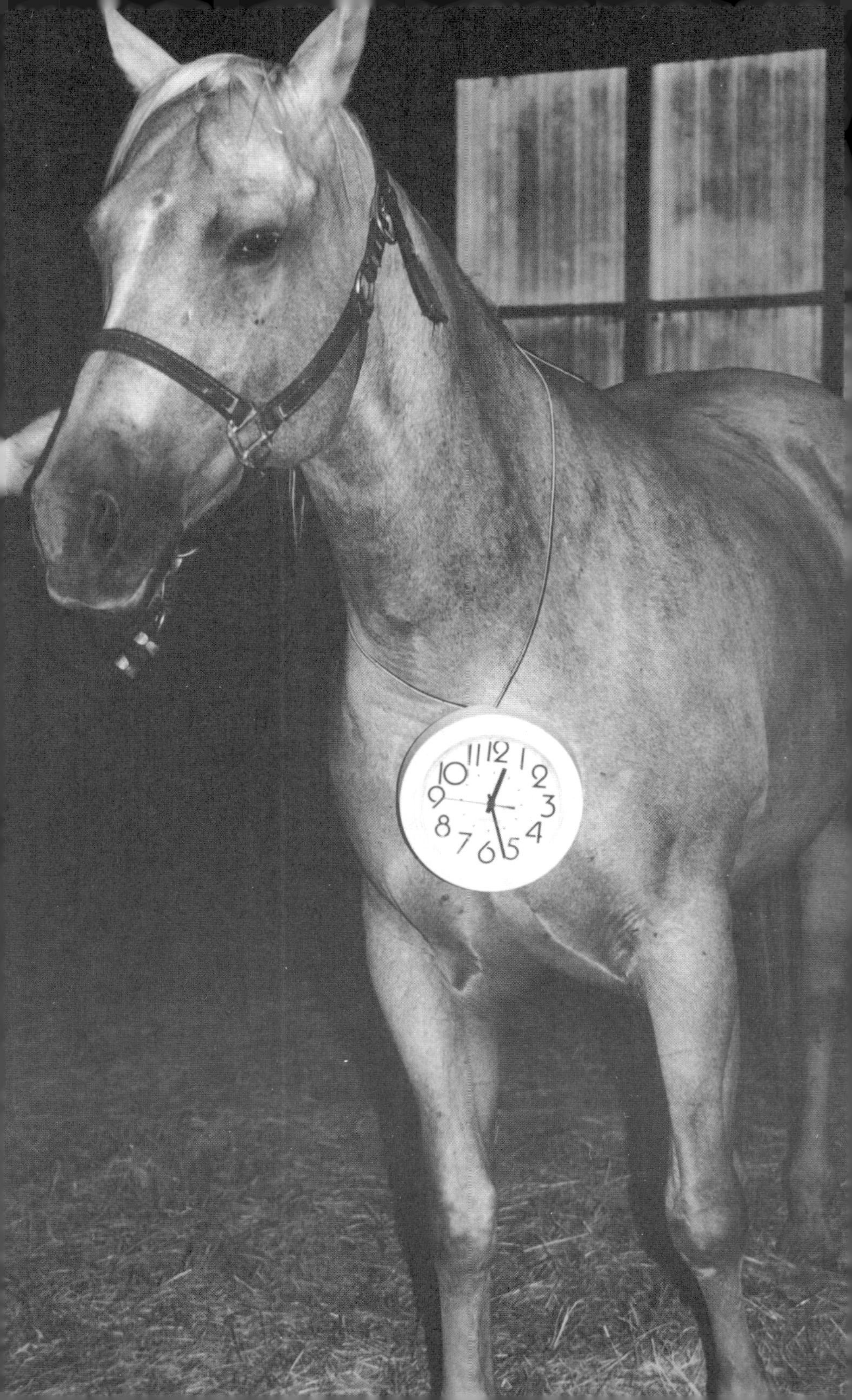
12
1
2
3
4
5
6
7
8
9
10
11

Chapter One

"Horse Time"

I know you are familiar with time. There is "Work Time", "Part Time", and "No Time". Also "Dinner Time", "Nap Time", and my all time favorite "Snack Time!!" ☺ Now let's get down to what you really want to learn. That is "Horse Time".

Horse time does not recognize any of the most important issues or situations in our world. Horse time just knows the world of the horse. The world of the horse is a lot subtler and less crowded with unimportant issues.

Let me discuss what is unimportant to a horse: jobs or school schedules and commitments, Little League or soccer games or practices, a special one-day sale at the shopping mall, birthdays, weddings or the Super Bowl. By now you can read where I'm going with Horse Time.

Horses do know and accept food, water, shelter, pain, pleasure and fear. Not to mention, anger and play, along with learning and work as the key elements to their world. This all adds up to Horse Time. Horse Time is what you really should be thinking of when you are around horses.

In some respects, horses really have a better life than we do. Horses can take as long as they want to eat a meal, they decide where and when to sleep, and they have virtually all day every day to accomplish the most fundamental of tasks. Eat, sleep and be merry really is what being a horse is all about.

Let me now show you just how fragile horse time can be. It is so easy to hurt a horse or prolong an injury just because of not understanding horse time. Very rarely does horse time and our sense of time ever match up completely. This is best illustrated in the following way.

For the purpose of explaining this, we will start with a barrel racing horse. Unfortunately, the horse has a minor bruised foot. No abscess danger, just a little rest and foot soaking with Epson salts. The horse doctor has told you to not ride the horse for about a week – no lunging, no hot walkers, and no stalls. Just turn the horse out for rest and healing. A simple little horsey vacation. Rest and recuperation: no big deal, no fuss, no muss!! This just can't be happening; there is the Wednesday Night Round-Up Club Barrel race and the Saturday Rodeo which you have already paid entrance fees for. Oh yes, non-refundable rodeo fees. The veterinarian just doesn't understand; you really have to use your horse! This is the only horse you have.

So you talk to a friend and decide to medicate the horse, possibly call out the horseshoer and inquire about a different shoe or pad or something, just anything so you can ride your horse. The horse cannot spend a whole week in a pasture just hanging around. Besides you saw him trotting around just today; he is probably ok.

The veterinarian is obviously just playing it too safe. The doctor does not know the whole story. You need to show up at these events. The vet stated it was minor. Your horse will just have to suck it up and get through it. Besides the arenas are soft dirt. It's only a couple of go arounds. So you do whatever you have to. You ride, haul and work that sore foot. The horse comes up lame; the bruise is twice as bad now. You have just created a three-legged horse. It's bad and getting worse. The shoe is off; the foot is bandaged; the horse is under daily care for weeks, not to mention your time doctoring the foot, the vet bills, the farrier bills and the look in your horse doctor's eyes when you explain what occurred.

What happened was a perfect example of horse time versus your time. The overwhelming majority of joint problems, muscle, and tendon strains all occur due to this situation - Not allowing a horse to heal on their schedule, not yours. Horses need what they need. Your priorities are not theirs; your goals, dreams, desires mean nothing to a horse. They can't understand them. Plus if they could, a horse could care less. Your time is your time. I'm writing about "Horse Time".

Take for instance a horse's hoof that is cracked or chipped. The foot really cannot hold a shoe securely for any reasonable length of time. You basically demand that your farrier or horseshoer put a shoe on the foot. They explain that it is not the best idea. Yet guess what? Your time takes command over horse time. The shoe pulls off and takes twice as much hoof wall with it. Nice going. Now it will be months versus weeks before that foot will hold a shoe.

I could easily take up this entire book with story after story of horse time being disrupted by our time. I don't think that is necessary. By now, hopefully, you understand just how important horse time is. Throughout this book you will come across horse time issues. Reflect back to the first chapter of this book.

Compromising always sounds like a failure or that you are giving up. Well, in everyday life maybe sometimes it can. As far as your horse is concerned, compromise is key to good horsemanship.

What you want out of your horse, or the duties and the performance level you demand from the horse, may not be possible or even just plain ridiculous. Yes, it could be just too far-fetched a demand.

I realize the cost and day-to-day work involved in owning a horse can be staggering. You want not just to enjoy your horse, but also to use the animal in any manner you choose. Although you deserve that right, your horse may not be capable of doing what you want.

Chronic injuries, temporary lameness, maybe a training problem, or something as common as old age creeping in: horses' life and physical abilities change often and very fast. It's these changes that we as riders tend to overlook or try to stop with drugs, corrective shoeing – anything or everything in our power to stop it. I really don't know how many times over the years I've been called on for advice on a screwed up horse situation. Did you know most veterinarians figure that the average horse owner will only do 50% or less of what the doctor has advised? Farriers and horseshoers figure their advice is probably around 25% to 30%. As far as feeding directions on feedbags, using supplements and hoof treatments, well who knows, but I bet it isn't any better either.

Why is this? How many times does someone read his or her vehicle's owner's manual? Or the directions before assembling anything? We like short cuts. We also don't like to compromise on any level. With your horse, compromise is everything. The more you listen to your horse's health care givers, the more you realize and respect your horse's limitations, the less money you will spend on vet bills. Also the more consistent and high quality of performance you will get out of your horses. The performance may not be what you originally wanted. It may not be what you require. Yet it is what is best for the horse.

Without a doubt, when you choose what is best for the horse, things go smoother and usually are less expensive as well. Yet choosing what is best is a tough choice to make. It means setting aside your personal desire to ride and use the horse. This could mean a loss of victory at a roping or barrel race or possibly a loss of points or association standing at a horse show. Missing a play day or trail ride could easily happen too. If you compromise just a little, sometimes this is all that it takes. Maybe that means a chance to heal up properly. Just a little rest and vacation. Also it could be a realization that your horse cannot perform the way you need it to.

Whatever the situation is, just about 100% of the time we make the horse compromise to our demands. The key words being "our demands". Your horse has no logical or rational way to justify your needs. The last thing I have to say on this matter is when dealing with a horse, remember you're not on your time; you're on horse time!!!

I have yet to come across any topic or situation regarding a horse's interaction with human beings that does not start or finish with great emphasis on the direct conflict of our time and horse time. It is the fundamental reason why out of all the subjects I could have written about, the very first chapter of the very first book I have ever written is on "Horse Time".

Chapter Two

"Inadvertent Learning"

Exactly what is inadvertent learning and what does it have to do with your horses? As hard as it may be to believe, the largest amount of information and training your horse absorbs is through inadvertent learning situations. First of all, let's cover what is classified as inadvertent learning. The word "inadvertent" means to overlook, absorb by accident, an oversight, something that goes unnoticed. Now for the word "learning" – to educate, to take information, gain experience and knowledge. Now "inadvertent learning" put into horse technology means your horse learns habits good or bad without you knowing it or wanting it to occur.

Horses have an amazing ability to pick-up on the subtlest of situations. I've always said a horse is like a ball on a pool table. For every action there is a reaction. Virtually everything a horse does good or bad is in conjunction with some other response. I will give you some of the more common inadvertent learning situations. Also I will show the learned behavior and the consequences of this theory. For every action there is a reaction.

1. Situation: Your horse is being trimmed and shod. Every time the horse moves a little the horseshoer jabs the horse in the side with a tool. This happens just about every time the horse needs its feet worked on.

   Inadvertent Learning: After a couple of times, the horse starts to associate the pain and mental stress of the jabs in the side with having its hooves worked on. Result - the horse becomes increasingly resistant to trimming and shoeing.

2. Situation: When a horse acts up or does not act properly. It is slapped on the side of the head.

   Inadvertent Learning: Eventually after being hit enough times any hand motion to the horse's head is considered a harmful experience. Result – your basic head-shy horse.

3. Situation: Every time you mount your horse you accidentally kick the horse in the side.

   Inadvertent Learning: The horse associates being mounted with a sharp unexplained kick in the ribs. Result – a horse that tends to move away when being mounted by a rider.

4. Situation: While riding your horse, your feet constantly make contact with the side of the horse. As rhythmic as the gait of the horse, your feet bang against the horse.

   Inadvertent Learning: This horse will grow increasing impatient while being ridden. Also this situation virtually destroys any leg or foot cues while riding.

5. Situation: Every time you do chores you make the same noises and make the same movements with your body.

   Inadvertent Learning: No matter what time of day it is, your horses always know when they are going to be fed.

Not all inadvertent learning is bad. Some may be so insignificant, the cause or the effect may not matter. There are a lot of situations I did not list that are very common, even a little humorous. How about a horse that knows the vet truck and refuses to be caught or it knows you're going to bathe it, worm it, and tack it up. Well, you get the idea. Now I want to discuss the deeper, more serious side of inadvertent learning.

Many, many horses live their lives in a state of confusion or panic due to inadvertent learning. Horses are no different than we are when it comes to pain and injuries. The pain and injury that involves the body can heal and be virtually forgotten. The pain and injury to the mind may never heal or never be forgotten. Mental stress and mental dysfunction can be very common place due to inadvertent learning.

Are there horses so traumatized by experiences or by living in an environment that is unfriendly to a horse so affected that they will never get over it and past it? Absolutely. There are countless numbers of horses in this situation. These horses do not respond to the average everyday way of working and training a horse. However, you can make good solid progress by first establishing what exactly is the action that caused the reaction.

Before you can retrain or help the horse with whatever the problem is, you must have an excellent understanding of what happened and how long it occurred to the horse. In other words, don't concern yourself at first with the reaction; dedicate your energies to fully understanding the initial action.

It is similar to being a detective examining a crime scene. Before you can clean up the mess, you must first carefully look at not what happened but why did it happen. Cause and effect; action and reaction; start and finish; your time and then horse time. Inadvertent learning is an everyday occurrence to a horse. You being the horse owner must decide if anything should be done. Please make no mistake about this subject matter. Horses in a pasture, horses in a boarding facility or horses running wild in a herd out west; inadvertent learning is taking place.

DANGER

Chapter Three

"Quicksand"

At a first glance at this chapter's title, you may be thinking of an old movie where some poor fool walks right into quicksand and disappears screaming for help. Well as melodramatic as that sounds, I'm going to talk about another type of quicksand. This quicksand is just as sneaky and can have just as dangerous a result.

You are doing what you love most, spending time with your horse. Things are going along just fine; then a little something changes. It throws you off for a second, yet you seem to be dealing with it. Then again, another problem, then another and another. Now no matter what you do or try, things are going badly. No answers, no progress. Your troubles are dragging you down. You're now in over your head just like "quicksand".

Medical emergencies like colic and founder can put you in this position. The ability to recognize that your horse needs professional help can be the difference between life and death. Pride and peer pressure for an owner have been the downfall of many a horse.

Quicksand can occur in a moment's notice on a trail ride, even in a crowded arena. Quicksand can suck you under. The ability to ask for help and assistance with your horse is not a sign of weakness or stupidity. The realization that no one person can know everything about a horse does not insult you; it does just the opposite. It shows you have the ability to do what is best for the horse. It also opens up the door to experience and new knowledge. Quicksand does not just apply to medical emergencies.

Quicksand can apply to training a horse, to horse barns and to horse trailers. Anytime, anything involving your horse can lead to a spiraling, never-ending disaster of a situation. I really hate to admit that no matter how prepared and how careful you are, you can still walk right into quicksand. No one can ever fully protect themselves and their horse from quicksand.

Yet you can keep it in your mind. Be on the lookout for a situation that can get out of control in a hurry. Try not to take risks with your horses. Also try to avoid or at least distance yourself from other horse owners and horses that have no conception of quicksand and its dangerous potential.

I am very fortunate to be in a position to learn and improve my horse knowledge and horse business experiences on a daily basis. The radio show, the horse clinics and the articles I have written, along with this book, plus the invaluable resource of my clients whose horses I work on every day: this is a classroom of immense proportions and uncountable information. Yet I have learned more from my mistakes than I have from my successes.

Although for one cowboy I have achieved a small amount of success in my chosen field, I have gotten more satisfaction and peace of mind from my ability to correct a mistake and recognize quicksand when I come close to it. Like most of you, I realize that in the majority of life nothing is free and nothing comes easy. Quicksand without a doubt is a very dangerous situation for you and your horse. Not recognizing or acknowledging the existence or potential of quicksand is much, much worse.

Learning new things, spending hours and hours with your horses, taking the good and bad information from other horse owners, always keeping an open mind and a keen, sharp eye focused on the unexplained and unexpected is the safest path around quicksand. As a final thought to this chapter, a human being's time has a lot more quicksand than does horse time!!

Chapter Four

"Chainsaws, Toaster Ovens and Other Useless Tack"

I'm always learning, with having a syndicated radio show about horses and putting on 30 to 40 horse clinics a year. I'm never out of energy when it comes to educating myself. Nobody knows everything, but I'm giving it a real big shot. I've got a long ways to go.

It's always interesting to me - the bizarre stuff, junk, crap (or whatever you wanna call it) people use to groom or train their horses. Come on ladies and gentleman, you know darn well who you are. Using an automobile window squeegee as a sweat scraper! I saw a toilet converted into a grain feeder. Oh, I almost forgot the time I saw a guy trimming his horse's feet with an electric sander. A lady once was trimming her horse's ears with a man's electric razor! How about this one: training a horse to drive by tying ropes around the horn of the saddle and pulling around an old dead riding lawnmower.

One of my all-time favorites has got to be the time I showed up at this guy's barn to check on a horse with a bowed tendon. I'm examining the horse cross-tied in the alley of the barn.

The horse owner is pleased that the horse is ready to ride. He starts telling me about going on an out-of-state hunting trip up north. Going elk hunting I believe. We're talking and all of a sudden out of a box stall jumps his wife, growling and pawing at the horse on her hands and knees. Okay, picture all this going on, not to mention the old brown carpet she has duct-taped to her back. Shag, I believe. I about jumped out of my skin, and of course the horse just totally blows up, breaks the cross ties and runs like the devil is chasing him. Well, more accurately, a woman with brown carpet duct-taped to her back!

I took a few minutes to calm down, gather my thoughts, basically just to be sure what happened really happened. I helped the husband gather the horse up while the wife took off the carpet. I then look at both of them and asked, "What in the name of good gravy are you two doing here?" The husband replies, "The hunting guide said we could come across a bear, so we thought we would pull surprise attacks on the horse around home, so he would get use to bears." I reply, "Who decided who gets to be the bear?" They look at me, and then each other like I just asked a strange question. The wife replies, "BJ, we take turns. Don't you know they have male and female bears!!!"

After a short while of pondering how many ways I could possibly respond to such an incident and conversation, I politely smiled and said, "Bears. You're right, what was I thinking"! A few months went by and I learned the husband was planning a fishing trip. As they were describing the fishing trip, as hard as I tried I couldn't help noticing the above ground pool. The ground was all tore up around the pool border. Instantly the famous music from those shark movies entered my mind. Da..... Da......Don. Don. Don. Don. It took all my inner strength not to go over to the pool and ask, "Who plays the shark?" Then I realized what a stupid question. "BJ, they do have male and female sharks."

Folks nobody – and I mean nobody – takes greater pleasure in learning new horse tips than I do. Like most things in life, there is a right way and then there is bear attack training!

P.S. I believe shark attack training is more expensive than regular riding lessons.

The moral of this story is try to use horse equipment and horse grooming tools properly. I love a good imagination as much as the next person. But reality is important when working with horses!!! ☺

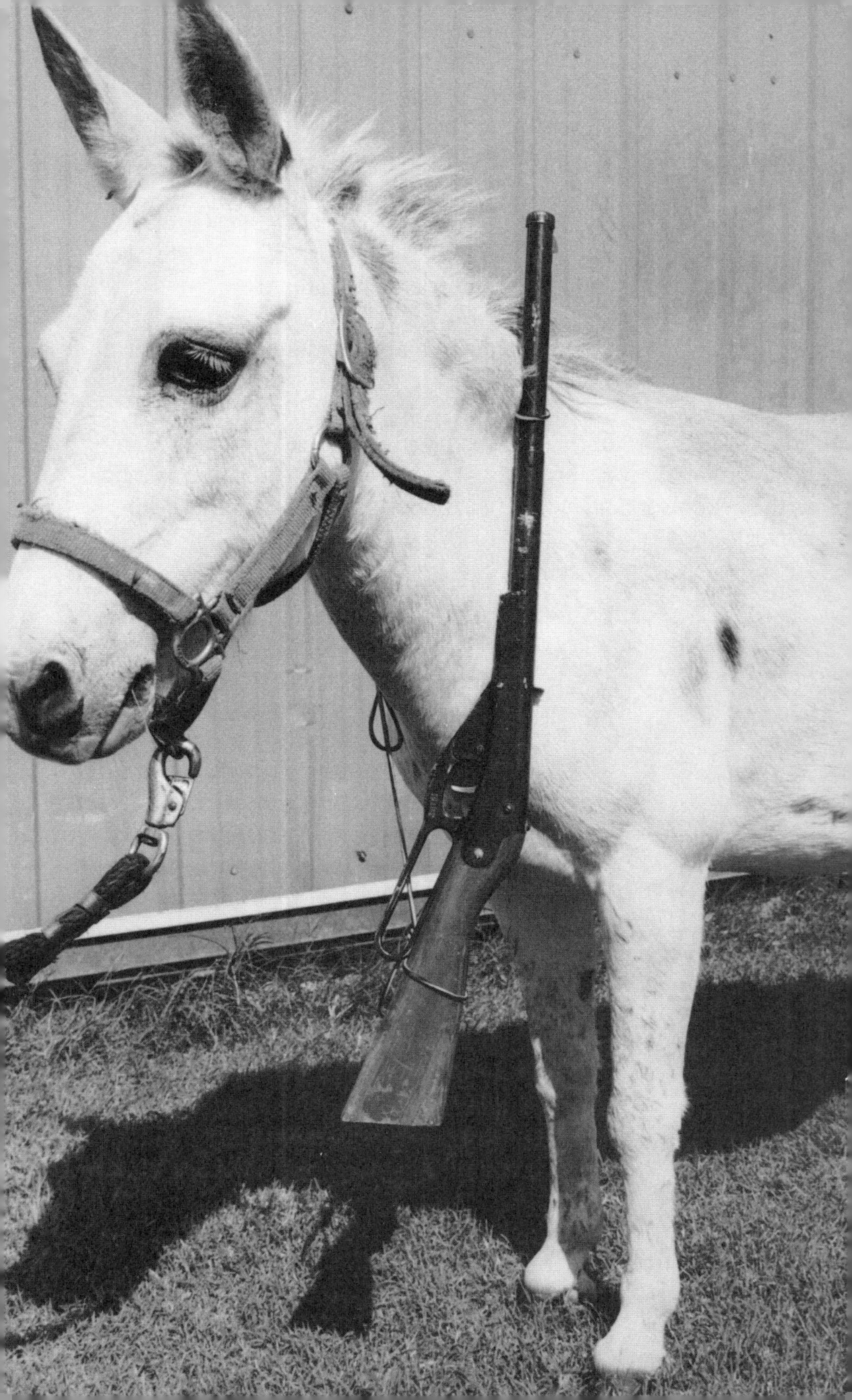

Chapter Five

"Clear and Present Danger"

"Clear and present danger" sounds very ominous, like a Hollywood movie. In this case, it states a real fact. Clear and present danger is what you're involved with when around a horse. You would think that being around a horse on a regular basis gives you confidence in how to handle a horse safely. The horse or horses are accustomed to you and you to them. We tend to think that day-to-day contact with a horse gives us a certain familiarity and confidence that will guard us against getting injured.

Unfortunately it is the day-to-day confidence and contact that actually puts us at risk. It is a real risk, ladies and gentlemen, a risk you all too often take for granted. If I sound serious - with just a touch of drama in my pen - it stems from years of experience hearing about or seeing so many horse-rider accidents, accidents that should have never happened at all. There is a mathematical equation for any and all things, a flip of a coin - heads or tails. The law of averages and possibilities can enlighten us all as to future events yet to be part of our daily lives.

It's the risk factor that teeter-totters the odds against us. The more hours you spend with a horse, the higher the odds are that you may be kicked, stepped on or bucked off. My personal favorite is the almighty bite on the back or on the upper arm.

These situations and, unfortunately for all horse owners even more disastrous situations, have a great possibility of occurring. Can you test fate? Can you change that mystical mathematical equation? Can you control that flip of the coin? How much can you avoid that horse-related injury? The quicksand theory!!

Well folks actually you may have more power than you may know. That's right; you can control the chances of being injured. Of course you can't have total 100% safety. Yet you can really diminish the chances of serious injury to yourself.

All horses kick, all horses bite, all horses buck and all horses rear up. Any horse can panic and spook. Let me repeat these two sentences again.

1. All horses kick, all horses bite, all horses buck and all horses rear up.

2. Any horse can panic and spook.

By the way, all or any one of these wonderful characteristics can pop up at a moment's notice, much faster than most riders can adapt. You have to stay alert and have a healthy respect for horses to stay out of danger.

So when riding, use a helmet if possible and don't ride alone; wear proper shoes and clothing and keep a sharp eye on any horse that is near you for any fast reaction to a kick, buck or bite.

Practice good old common-sense safety when around any horse, even those old tired pushbutton baby-sitters. Even they can really set a person on their ear. By the way, that's literally "on your ear." So watch where you walk. Think about how you're handling the horse, and then make any adjustments necessary to keep yourself safe and healthy.

As I close this particular chapter, I'd like you to think about something. Do you know why most accidents happen in and around the home? It's not due to any real clear and present danger. It's because folks are just so relaxed and overconfident in their own home environment. Confidence and day-to-day again!!

Ride a lot and ride smart!!

MINE
YOURS

Chapter Six

"Break-ups, Divorce and Your Horse"

You're probably wondering what break-ups and divorces have to do with horses. Unfortunately more than I care to write about. Everybody has heard the term "pawn" – using the children against Mom or Dad in a divorce. If you think that sounds tough, horses suffer as much or more when hearts are breaking and battle lines become drawn. When a relationship breaks up, no matter how mature or adult people try to be, it seems sooner or later the battle starts: sometimes hiding or destroying horse registration papers; not doing chores or being negligent in the horse's care; even sometimes being harsh or physically cruel to the horse; taking horses away and hiding them; separating pasture mates that have been together for years; weaning foals early; selling off horses on the sly; not knowing or caring about what type of new home the horse will have.

Consider also not paying vets, farriers and feed stores; stopping medical treatments; cutting off or switching a horse's diet; taking tack away. Just imagine the poor horse that has become used to a certain bit, bridle or saddle. All of a sudden this horse has to adjust to strange tack, blankets, pads, grooming supplies, buckets and feeders. Remember folks, horses like habits in daily routines. Any change should be done gradually, or not at all!!

At least a couple of times a month I come across a bad break-up. I try to give this advice to the parties concerned, yet rarely does anyone listen. I would never try to give relationship advice. I also stand hard and fast in my belief that each horse owner has the right to make any and all decisions concerning his own personal livestock, excluding animal abuse or cruelty as described in all state statues and laws (not to mention common sense).

Try to realize horses have absolutely no ability to understand divorce and or break-ups. Horses just know food, water, shelter, their tack, and most importantly who feeds them and where home is - safe, secure and familiar home. One of the most important terms that can be used to describe proper horse care has got to be consistency; that is, everyday safety and routine.

Ladies and gentleman life is long and can change as fast and swift as the wind. You never can predict your future. However, you can help assure your horse's future with proper health care and a safe home. As I always say at the end of every radio show, also at the end of each speaking engagement I do, "Take care of yourself and take care of your horses."

Chapter Seven

"Just a Little Horsey"

A lot of you old timers – yes, old timers – know what "a little horsey" means. It doesn't mean you're full of horse crap. It's actually a very good compliment. Being "horsey" means you have a good solid grasp on the fundamentals of horsemanship. There is a lot more to horsemanship than sitting a good saddle. Nowadays, unfortunately, riding a horse has become the benchmark for horse skills.

Whether you are financially secure and have someone doing your chores and grooming your horses or you just have one horse in your back yard and it's a struggle to keep it, your economic situation has nothing to do with knowing and using horsemanship skills. What do you need to know?

First let's cover the basics. How much water does your horse drink? How much hay does each horse eat? Which is the alpha horse? Do you personally watch each horse get its shots and worming, getting its feet trimmed and being shod? What are the basic points on a saddle? Average healthy horse's temperature? Emergency first-aid procedures?

Then there's hitching, loading, pulling a horse trailer in any weather, anytime. Spotting a lame horse, noticing a snotty nose or a casual coughing in the barn. Knowing the difference between a good deal on a used saddle or getting taken. Being horsey isn't something you learn from a book. There isn't any test or graduation with a certificate of achievement. Very rarely can you learn it from someone else. It would be great if you could.

Becoming horsey takes time. I mean years, being and doing. Anyone can own a horse. Also anyone can be taught to stay in the saddle. But only through day-to-day horse maintenance can you really learn horses – things such as your horse's traits, tendencies, mood swings, and everyday horse care.

Is not being horsey a dig at you personally? An insult to your reputation? Absolutely not. It's a goal to achieve, something to strive for. Will you know when you reach your goal? No, probably not. You'll be too busy cleaning, fixing, grooming and working horses.

Being horsey is not a title or ribbon or trophy to hold and display proudly. It's the inner peace that comes with the knowledge and skills only trial and error can achieve. So be patient. Soak up all you can. Don't avoid a task or chore; delegating chores or responsibilities isn't the answer.

Wearing out a pitchfork. Knowing each barn cat's meow. Listening to a truck coming up to the barn and knowing if it's the farrier or veterinarian. Going to the feed store so often you know everything in that store. Having no idea what is the hottest TV show on. Could care less what the weather report says. You have a lot to do with your horses tomorrow. Is that being horsey? Let's say it's a good solid start.

Chapter Eight

"The Greatest Act of Love – Euthanasia"

Sooner or later in your lifetime, someone you know and care about dies. Death is difficult to accept. Death is even tougher when you must decide that it's time to end a beloved animal's life. When the decision is up to you, you may face many hard choices. Unfortunately, if you own a horse, whether just for a month or all your life, it's something to keep in mind.

"Putting a horse down" is the common barn term for humanely ending a horse's life. Medical reasons, safety reasons, even economic situations can be the deciding factor for euthanasia. The choice or decision to put down a horse has to be a careful one. Ultimately, what is best for the horse? Consider the situation at hand. There are a lot of reasons for putting an animal to rest. The following is a list of the more common situations and conditions that can put you, the horse owner, in this very difficult place.

1. <u>Chronic Lameness:</u> A horse is in so much pain it cannot sustain its own weight. Damage so severe, no healing will take place.

2. Incurable, transmittable disease: The first thing that comes to mind is a positive Coggins test. A horse with E.I.A. or Equine Infectious Anemia should be put down so as not to pass the disease to other horses.

3. Inoperable colic: Colic is responsible for killing a large number of horses. Colic in its severe state cannot be cured.

4. New Born Foals: Foals born with severe defects or problems. Enough said.

5. Old age: Can't walk, can't eat, poor overall health. It's the nature of things.

6. Severe traumatic injuries: Every vet dreads that middle of the night call. Accidental injuries or an injury too large. Maybe medical attention cannot be given in time. The horse is just hanging on.

7. Violent Behavior: Fortunately rarely seen or experienced by the average horse owner, but there are violent, dangerous horses. These "man killers" are usually spotted at an early age. They become a danger to themselves as well as others.

8. Money: Yes money. Sometimes a horse is so badly hurt or its health is so poor that maintaining it can cost three to five times its value. Usually the horse cannot be used. Most horse owners do not have the

luxury to afford spending thousands on a horse that will never be ridden. Also the horse has no "sale value" at all.

9. Agony: Suffering, pain, sadness. The horse is experiencing shear torture just surviving another day. Quality of life is dismal.

Each and every case is unique. Although you may be able to compare your horse's situation to a similar one, each horse has to be evaluated on its own merits. Some horses can survive and heal, while others will not. Some horses are tougher, even healthier, and can weather the storm, whereas other horses give up and can't go on.

"Putting a horse down." This subject is so difficult. Most magazines and books won't cover it in any detail. My radio show is the only show that has ever discussed this in exact detail, right down to how the drugs work and how the horse reacts as it passes away. Personally, folks, I feel if more people understood what a horse goes through, and how easy it is to lose a horse to accidents or diseases, we would have fewer dead horses. Now let's talk logic. Cold hard facts. These are questions to consider in a conversation with your veterinarian and a close friend.

1. Will the horse heal enough to become useful?

2. Has the horse lost hope? Has it given up? Does it desire to survive?

3. Is special care required? Can you provide it?

4. Can you afford this care?

5. How much pain and suffering can you – yes, you – watch your horse go through?

6. Are there any other choices, decisions, and alternatives to this situation?

7. What are insurance considerations, if this horse has coverage? Your insurance company may want a say in the horse's future.

8. What will be the physical ramifications of keeping this horse on pain medication for a long time – possibly the rest of its life?

The relationship that you have with your horse doctor is never more crucial than now. You need to trust his/her opinion. I've never met a vet who enjoyed this process any more than the horse owner. Only under extreme cases will a vet put a horse down without the owner's permission or knowledge: an incident such as hunting accidents; the shooting of horse; a car-horse accident; or an accident on the racetrack. In these cases, speed is the humane answer.

However, a vet may decide against this. The doctor also follows his/her own heart. The doctor may feel the animal does not need to be put down, or if the horse owner is causing suffering, the vet may just stop treatments and walk away from a bad situation. All of these factors make the owner-doctor relationship so important.

Let's move on to planning horse euthanasia and getting it done. Sorry, but you really need to know all of this.

1. Make your decision and decide where is the best place would be for this to occur.

2. Let boarding stables know your plans. Also let any other stable mates know as well. This is not the type of thing you want to surprise anyone with.

3. Before the procedure, make arrangements for disposal or burying of the animal.

4. Contact your insurance company about your plan and intentions. The company will want a report from the vet.

5. Probably the biggest thing to consider - you may not want to be there during the procedure. A lot of folks would rather the last memory be of what the horse was, not of what the horse has become. A friend can take your place during the procedure.

6. Have the doctor go into detail on what to expect, such as how the horse will react.

7. Take time to be sad. Grieve. Cry. Don't suck it up. Let it out. There's nothing wrong in this. You're losing a big part of your day-to-day life.

We grow to love our horses. We give them names, birthday and Christmas gifts. A lot of us would rather spend time with our horses than people. As compassionate as we are, we tend to assign our horses more personality and a higher consciousness than they actually have. It's this personality and close bond that makes putting a horse down so hard. I believe it should be hard. Hard enough to keep our horses safe, happy and healthy, and to cause us to make good, sound choices every day for our horses' overall well being.

Ladies and gentlemen, I personally observe two to four horses every month being put down. In some cases, it's time – nature has run its course. Unfortunately, most of the time, a tragic mistake or accident occurred. I never take it well. I probably never will. I do believe though that whatever you choose to call it – mercy, kindness, deciding to give an animal lasting peace, a permanent end to pain and suffering - putting a stop to a life of just hanging on and surviving – in this horseman's heart, it is the greatest act of love. All my best to you and your horse.

I would like to say that this particular chapter is in honor of my pony "Dave". Dave had a wonderful and varied life. I made sure in his last years he had not one bad day. I put Dave down in February of 2002. The sun was out and it was a mild, peaceful day. I dug his grave by hand; I needed the time to really consider what I was about to do. Every shovel of dirt I pitched out of his final stall gave me that needed time to reflect on his life.

DAVE WAS 46 YEARS OLD, AND IT IS THE LAST TIME IN MY LIFE I WILL BE ABLE TO OWN A HORSE OLDER THAN MYSELF...

A PONY AS OLD AS DAVE, DESERVES THE FINAL WORDS HAND WRITTEN... HIS PASSING WAS SWIFT AND PEACEFUL...

Chapter Nine

"Stealing Horses"

You arrive to take your horse out for the day, but something is terribly wrong. The lock on the front gate has been cut off and is lying in the dirt in front of the gate opening. Fresh tire tracks leading away from the scene prove this was no escape. Do you know what to do or who to call?

Each year, about 60,000 horses are stolen by bands of thieves who often work in one area or state. When authorities notice a sharp rise in horse thefts and start to close in, the thieves then move into another area. It's quick, profitable and dirty work, because most horse thieves swiftly unload their stolen horses at auctions within the first day or two. And as most horse owners know, some horses sold at auctions may end up in the loving care of a new horse owner, but most go to slaughter.

Experts say that actions you take in the first 24 to 48 hours are crucial if you hope to recover your stolen animal. To increase your chances of recovering your horse, you must act quickly and effectively. The more quickly you act after your horse is discovered missing, the better your chances are of recovering it.

STEPS TO TAKE

Act swiftly. When you find your horse is missing, don't let a nanosecond go by. In a missing livestock incident, the more time that goes by after the first 24 hours, the less likely your chances are of ever seeing that animal again. Usually, when horses are stolen, the owners are so shocked that they don't know what hit them. By the time they get moving it's too late. Part of knowing how quickly to move is knowing what to do beforehand. You should have documents, photographs and contact numbers up-to-date and readily available, but even if you don't, you can still launch an effective search.

Escaped, stolen or mixed up? When your horse is not where it should be, first check for signs of theft or escape. Evidence of theft includes forced entry, fresh tire tracks, cut or broken fencing, missing halters or tack, open barn doors or gates, illuminated lights, missing trailers, or anything unusual in the barn or on the grounds.

If your horse has escaped on its own, you may see bite, scrape or kick marks on its stall door, or spilled feed or grain if the horse helped itself to your feed storage on its way out. In the pasture, inspect fencing for signs of a horse climbing out, knocking something over to get though or strands of mane caught on the fence. If you have tufts of mane hanging off a fence, it's usually because a horse has forced himself on his own through this space. If someone is stealing a horse, he's not going to pick the hardest place to get through. Because horses usually tend to follow one another, other horses from the same pasture outside the fence or hanging around the opening may indicate the missing horse left on its own.

Make sure there isn't any logical reason for the horse to be missing. Perhaps a barn worker, family member or friend moved the horses to another place on the property or is out riding it. Perhaps an emergency prompted someone to haul the horse to the vet. In the case of a large breeding farm, check that there wasn't some mix-up on the time a mare was to be picked up and taken away for breeding.

Work with the law. If your horse may have been stolen, report the theft immediately to local authorities such as the city police or the sheriff's department. Ask them to come out and look at the crime scene, take photographs and open an investigation. Unfortunately, horse theft is not a high priority item, so authorities might not come out right away, or at all, particularly if your horse has no identifying marks. Regardless, insist that authorities take a case report, even if only over the phone including a description of your horse, noting any brands, freeze marks, scars or other markings. You may need the report later for insurance purposes, court cases, or to document to other agencies that a theft transpired.

While waiting for the authorities to arrive, try to put together a time frame for the crime. Determine the last time you or others saw your horse. Ask barn workers, family members, neighbors and others who might have been around if they saw any suspicious vehicles in the vicinity, somebody stopping to pat or feed grass to the horses, or if the dogs made any unusual commotion to signal that someone is trespassing. If so, try to pinpoint the time of occurrence. Report your findings to the authorities.

Also, get out your bill of sale, veterinarian records, breed registration papers, and other documentation and color photographs that will help identify your horse. You have to be prepared so a non-horse person or an individual who has never laid eyes on your horse can identify it. If you have a plain brown mare, for example, pay particular attention to providing a complete description of her including scars, an odd marking, an odd hair swirl, any tattoos or brands or microchips.

Distribute flyers. After contacting the law, create flyers that contain a clear photo and a description of your horse – breed, color, distinguishing marks, age, size, weight, etc. – and phone numbers where you or others can be reached day or night if anyone has seen or has information about your missing horse.

Fax, express mail or hand deliver your flyers to slaughterhouses, rendering facilities, livestock sales and auctions, racetrack and rodeo managers, and middleman horse traders within a 500-600 mile radius. Also notify or send flyers to the State Department of Agriculture (Division of Animal Services), the state veterinarian, state cattleman's association, breed association, state horse council, equine veterinarians, neighbors and local farm or area equine newsletter. Some of theses folks may be able to assist with rescue and recovery. Post flyers at your local feed stores and tack shops as well.

Visit local livestock auctions. Compile a list of local auction houses, especially the ones that meat buyers attend. It is important to include private sale barns, but generally meat buyers aren't there and those horses aren't in danger of being sold for slaughter right from under your nose in 24 hours.

Find out when the auctions are held and look there yourself or send someone who knows the horse. Check the parking lot for private sales. Look in all trailers and holding pens. "Hot" horses show up in auction yards seconds before the actual sale begins.

Call slaughterhouses. Check in daily with the foreman and ask if there's a horse matching your description. A slaughter facility is more apt to look for an animal that carries identification. There is no way you can prove that any horse is your horse.

On the other hand, if the horse is micro chipped or has a brand or tattoo, you can call the slaughterhouse every day and request that they scan or inspect any horses that fit your horse's description.

Advertise. Place ads in the local and regional equine papers. Sometimes leads turn up.

If you find your horse, keep it under constant surveillance without being obvious or conspicuous. Don't try to reclaim the horse by yourself, as this could be dangerous. Summon the law enforcement authorities right away and let them take care of the recovery of your horse.

Nothing is worse than not knowing what's happened to your horse. If you suspect your horse has been stolen, enlist the help of friends and family immediately. The more you do as soon as you find out, the better your chances of recovering your horse.

PREPARATION AND PREVENTION

Being prepared for a theft and taking steps to minimize it and maximize recovery will go a long way toward protecting your horse and getting it back. Employ the following precautions:

1. Permanently identify your horse with a tattoo, brand or microchip. This proves ownership, aids in a horse's recovery and if visible, discourages thieves. Tattooing the upper lip of a horse is common in the horse racing industry. Branding horses, whether it is done with a traditional hot brand or now the more popular freeze branding technique is also a great identifier. Microchips are about the size of a large grain of rice. These have a numerical code inside them. Once the microchip is placed in the upper neck in the nucal tendon by the mane, it is there for life.

Microchip companies keep a record of each horse's microchip number and provide scanners for veterinarians and slaughterhouses to identify horses. The only down side to microchips is unless you know for sure that a particular horse is yours, it is unreasonable to assume every horse would be scanned looking for your microchip number. In some cases, however, the microchip is invaluable. Basically if your horse is stolen, let's say it is sold out of state and you find your horse, how do you prove it is your horse? The new owners could even have registration papers. Sure they're false papers, but you are the only one who knows this. As far as the new owners are concerned they bought this horse. It's not yours. That is when a microchip can get your horse back for you.

2. Take color photos of your horse, including head and full body shots (from both sides) Get photos with summer and winter coats, noting any distinguishing marks. If the animal has just one little piece of white on his coronary band, make sure the photo is large enough to see it. Choose five to seven inch photos with the horse filling out about 75 percent of the photo.

3. Keep photos, bills of sale, veterinary records, breed registration papers, and other documents together in an easily accessible notebook or folder.

4. Know what auctions operate within a 500-600 miles radius of your area, their addresses, phone, and fax numbers, and when they hold sales. You can usually find their information through the state department of agriculture or through the state veterinarian's office.

5. Remove halters from pastured horses. It is recommended that you don't leave them hanging near stall doors, but this is a choice to make. Many horse owners prefer to leave a halter by the horse's stall in case of emergency and will argue that most horse thieves come prepared with ropes and halters anyway.

6. Check your horses regularly, but deviate from the routine. Professional thieves will watch your place before they go onto your property, so vary your schedule. Day and night.

7. Install lighting with photo sensors or motion detectors around the barn or property gates, or place streetlights by pastures along the road.

8. If possible, replace wire fencing with board fencing. Wire fencing is easily cut. With board fencing, thieves have to find a gate.

9. Keep wire fencing in good repair and as sturdy as possible.

10. Use good gates, heavy chains and padlocks.

11. Use thoughtful landscaping. Plant trees and shrubs between the pasture and road to block visibility of the horses from the road, but keep bushes trimmed or absent around gates and barn doors.

12. Post no trespassing or warning signs.

13. Consider an alarm system. Keep dogs, or loud animals around to make noise when something is amiss. Set up a neighborhood crime watch program.

14. Observe the movements and behaviors of unfamiliar people. If they seem suspicious, report their actions to law enforcement authorities. If possible, videotape or photograph strangers that arrive at your barn, and record the license plates and types of vehicles. Many people casing your property may come out looking like a happy-go-lucky family with male, female, and a child or two. They may come in an old vehicle, but not necessarily a pickup, in the guise of wanting to pet or feed the pretty ponies. If you have recorded images or a license plate number, if shortly thereafter something is missing, you have leads to provide law enforcement agencies.

Unfortunately, someone reading this book will have a horse stolen sooner or later. A lot of times, random thefts of one or two horses are carried out by someone who knows the victims. Also, thieves know we all have jobs, school, family commitments; we only have so much "free" time to look for a missing horse. Horse thieves a lot of the time plan on this. They will steal a horse, hide it out for 2 to 3 weeks waiting for the victim to give up looking, then sell it at an auction, sometimes as close as 50 miles from where the horse was taken from.

Please don't give up. That is what the thief is counting on. Thieves hide horses in backyards, barns, and sheds, even a household garage. Sometimes the horse is kept in the trailer and is randomly moved from spot to spot. I always keep this perspective in mind. If a horse had no legs it would be close to the size of a big couch or piece of furniture. So you are looking for a piece of furniture that moves around on you while you're searching for it. This means look everywhere real close. Fortunately for a lot of horse owners who have the misfortune to have a horse stolen, the thieves are not the smartest people. The thieves will frequently hide the horse within minutes or just a few miles from the horse's home. Make sure to check the "local area" a lot and at least five to ten times over a week. Remember the big piece of furniture theory!!!

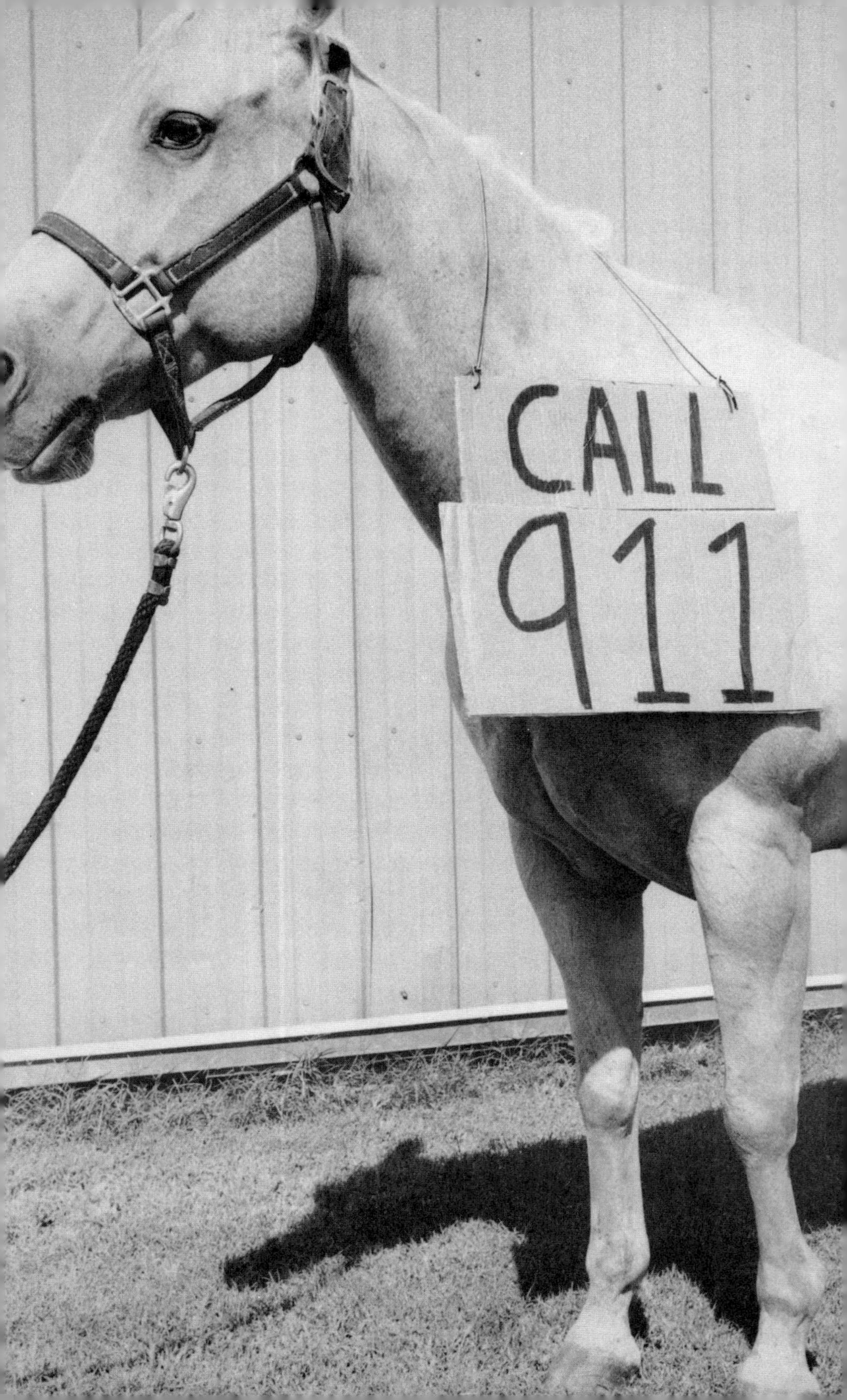
CALL
911

Chapter Ten

"Help! Police!"

You're probably wondering what the police, or more officially law enforcement, have to do with your horses. Well, actually a whole bunch. Unfortunately, nowadays a large number of horse owners will sometime in their lives have to dial 911 – either over stolen horses, tack theft, drunk drivers crashing through fences, or disputes with property owners. Even though horses are the number one pastime in the U.S., when it comes to laws regarding horses and enforcing them, your local officer that covers your area is basically clueless.

This is not criticism of those officers. It's just that horse crimes are few and far between, considering other crimes. Also, as far as most people are concerned a horse is like a pig, cow, dog or cat. A lot of folks don't appreciate how much time and money you've invested in your horse. Even though all states do have laws for boarding, selling, even transporting horses, sometimes the laws are general domestic livestock laws.

Since man-against-man crimes are a much greater priority, horse crimes often fall through the cracks. Sure, every once in a while you'll hear of somebody going to jail for theft, but almost never for cruelty or fraud. Most district attorneys just feel intimidated, not to mention inexperienced in livestock crimes. I've had judges and attorneys tell me a crime is a crime – no difference. I find in the "real" world, horse crimes are very low on the crime solving priority list.

Let's talk about why you would need to call the police. Stolen horses. Your saddles, tack gone. Horse trailer missing. Fences torn down by neighbors or someone driving off the road. A dispute with a horse boarder. Your horses get into trouble, out on the road or in the neighbor's yard. Stud on one side of a fence, mares on the other. Unfortunately, this book and my radio show cover too much of the country for me to get specific on your local laws. Yet, I can give you a lot of good tips on how to educate yourself on what is important to your particular area.

Your local library will have the laws, state and local books. Look up horse and livestock statutes. Also contact your local judge or district attorney about speaking in front of a group of horse owners about crimes and punishments. They will do this; they are in elected political positions. An audience, especially voters, will get their attention. Contact your police or county sheriff. Find out which officer or detective handles livestock investigations. They will be glad to talk to you. Public relations is important to all agencies.

A little forethought on your part, as the horse owner, can really help. How, you say? Well let's say you call 911 for any horse crime. The officer comes out, listens to what you say happened. Then depending on what actually occurred the officer may or may not take a report. If you know your local horse laws, this is a perfect time to speak up and let them know a small part of the law that possibly the officer hasn't ever been exposed to before.

Also, officers on a local level tend to deal with a lot of nitwits. Dealing with a well-informed person is sometimes a refreshing thing. This can motivate them to go that extra mile to help you out. I personally take care of an entire police department's mounted patrol unit and several state troopers' personal family horses. Also a couple of attorneys' horses as well. Before writing this chapter, I asked a lot of questions. To my amazement and to some of their surprise, I was more versed on livestock laws than they were. Don't consider what I'm saying as bragging. It proves my point. That's why I decided to tell you about this. Please take time to discuss this issue among your barn buddies. You'll find a lot of questions and not too many answers. Good luck and I wish the best for you. Hopefully this is information you will never use.

Chapter Eleven

"Pony Proofing"

Often times you go ride, do chores, or maybe show someone your horses only to see blood and even a lame horse. Somehow, someway unknown to you, your horse has injured itself. Maybe it's minor; all you need is a little home doctoring. Or you may have a situation on your hands that requires calling your horse doctor.

Can you stop all horse accidents? Put an end to all self-inflicted injuries? No, absolutely not. But you can drastically reduce the risk. I mean really make a positive difference in your horse's safety. You have probably heard the term "child proofing your house". Let's "pony proof" your barn and pastures.

Start by making sure horses can't get their legs caught in any gates, buildings, water tanks or feeders. Inspect all buildings for loose boards or tin. Look for nails, screws or bolts sticking out of walls and doors. Look at preventing fencepost impalements by purchasing T-post protectors. Remove any junk cars and farm equipment. Pastures can be littered with all kinds of junk. Make sure fences are safe – no sagging or loose wires or panels.

Try to fill in any holes or dug-out places with dirt. Gopher, fox, and rat holes are notorious for twisting fetlock joints. Try to eliminate hanging strings, rope or wires, so horses have nothing to poke into their eyes. Also remember to protect horses from fans, heaters, and extension cords – any type of electrical device or outlet.

Garden hoses seem safe until a horse gets its legs caught up in one. Lead ropes, halters, and all kinds of tack can cause problems as well. Secure all gates, doors, and windows. Make sure your horses only have access to what you want them to get into. Always make sure all feed rooms and hay storage areas are secure from horses. Most cases of colic and foundering happen when horses force their way into feed rooms.

Leaving horses unattended or turned out with tack on can lead to many safety issues, such as broken necks caused by halters and bridles catching on things. Ropes and reins dragging can wrap around horses' legs. Feed buckets and water buckets that are not properly mounted are also hazards, along with salt and mineral blocks that are placed in doorways or areas where horses may stumble over them.

I realize, and I hope you do too, that horses by nature are very curious animals. Horses also get bored easily and look for things to do. Horses enjoy playing and goofing around. It's this curiosity and playful way that lead a horse into trouble. Horses panic very fast. In this panic state, they run or back up or rear up.

Not to mention, horses do have little fights and battles with their pasture or barn mates. The safer and cleaner you keep your horse's barn, stall, paddock, and pasture the fewer the chances of injuries.

Like I mentioned before, you always "child-proof" your home. So please take the time to "pony-proof" your horse facilities. It's much quicker and cheaper to fix and clean a barn area than to spend time doctoring your horse or paying vet bills.

Chapter Twelve

"Riding Alone"

It really doesn't make much difference - whether you have been on a horse your whole life or if you just bought your first horse - one of the most relaxing and thought-provoking activities you can do is take your horse for a nice long ride by yourself. For a lot of us horse owners, we do our best heartfelt thinking on the back of a horse. Life at any moment can deal us a shocking blow, leaving us down in the dumps and totally discouraged. Once you touch that horse, look into those eyes, feel every breath, hear every sound, your troubles just take a back seat. Even if taking your mind off your troubles is a very temporary situation, it's still a wonderful moment in your day. The peacefulness and solitude of a nice private ride is almost indescribable. Granted, sometimes you are in a wonderful mood and going for a ride is a great, exhilarating experience. Not all private times are for serious contemplation.

Yet there is a perspective to riding alone: physical risk. That's right. Danger is always around the corner when riding a horse. Let's talk about the fact that any horse, anywhere, with any rider, can spook, buck, kick, panic and bolt at a moment's notice. Unfortunately, as hard as you try to de-spook your horse, something can always pop up out of nowhere to cause your horse to go crazy, even those over-trained, dead-broke, push-button kid horses. Every time you mount a horse you're at risk, no matter how many years of riding experience you have.

It is not a matter of **IF** something would happen; it's **WHEN** it happens that counts. We have all heard a story of someone having to walk back to the barn after being, shall I say, "unintentionally dismounted". It's so easy to be seriously hurt falling off a horse. It's tough to catch a horse, stop bleeding and keep from going into shock all at once.

Bottom line, folks. Riding alone is just plain reckless. I could go on and on about types of dangers and pitfalls in being alone. Remember that your personal safety is not the only issue. The horse's safety is also a major factor. Horses are by nature herd animals. Safety in numbers is a built-in genetic trait of the horse. A horse very rarely feels completely comfortable alone. Horses rely on each other for security, keeping their eyes peeled and ears rotating for danger. By taking a horse away from its home and riding off by yourself, you're placing your personal safety with the horse who by nature is uncomfortable in this position. Some horses handle it better than others do. All horses on some level feel nervous going out alone. That's one reason why some horses try to speed up on the way back to the barn. The horse knows what and where it has to be, to be safe from harm.

Now that I've made my case for the risks involved in riding alone, let's talk about what to do if you're going to do it anyway. Ladies and gentlemen, I've been in the horse business for a long time now. I've learned a lot. One of the most important things that I've learned is that people are going to do what they want no matter what they are told. The following is a list of what you should consider doing if you choose to ride alone.

1. Tell at least two people that you're going for a ride.

2. Make sure to leave information about the trail, a map or direction of travel. Do not deviate from this plan. It's much easier to find you if they know where you've gone.

3. Also, make known a time of departure and a time of return, plus, how long to wait before searching for you.

4. A cellular phone or two-way radio is always very handy.

5. Try to ride only in familiar places. Save the adventures for when you're riding with someone.

6. Take a halter and lead rope along in case your horse can not be ridden, or if you can only walk.

7. Know your horse. Avoid areas, animals, and situations that can spook your horse.

8. A riding helmet is needed. (Although I realize cowboy hats and baseball caps are so much more popular). Head injuries are very possible when falling from a horse.

9. Never, I repeat never, ride at night, or in a thunderstorm with lightning. Believe it or not, a lot of riders do this.

10. Please be careful. Be a smart rider.

Well, I hope I've given you food for thought. Reading this may help you or someone you know. I didn't cover the medical problems that can occur falling off a horse. It seems, unfortunately, most of the injuries are serious. Is everyone who reads this going to stop riding alone? No, of course not, but maybe you will just take a moment to think about safety.

Chapter Thirteen

"The Lab Rats"

This book would not be complete without something being mentioned about the "Lab Rats". Especially since the lab rats are more popular than I am. Every horse clinic I speak at or personal appearance I make, the lab rats always are one of the most inquired-about areas of my life. The lab rats are revered as some of the most talked about horses in the United States. Reporters, when interviewing me, always ask how the lab rats are doing. Never how I'm doing.

The feed and tack companies that know about my radio show and horse clinics always ask me if the lab rats have tried any of their new products or tack. Although I have been offered just about every type of feed, supplement, tack, shampoo, fly spray, or new improved hoof pick there is, everything that the lab rats use or try out is paid for by me. It is always purchased at full retail price, including tax. This insures an accurate and honest comparison of new products on the market. Plus, I know just what you folks deal with as well.

If by chance you are unfamiliar with my radio show or I have not had the honor and privilege of meeting you at a horse clinic, the lab rats are my personal horses. I love them dearly. Actually I'm looking at them out my living room window as I write this. I have learned so much of what I know about the world of horses from them. I have yet to meet any horse owner who does not try new things. Feeds, supplements, and grooming supplies, whatever the instance is, your horses are an invaluable resource for education.

My own horses have taught me unconditional love and a sense of companionship that is hard to explain. Oh sure, the lab rats are always being subjected to some off the wall experiment. New products or new training techniques go without saying. To their credit, the lab rats rarely show signs of dismay or confusion. Although the firecrackers and fruit pies did throw them off for a moment, it was a hearing/sense-of-smell test.

The term "Lab Rats" was given to my horses for their resilience and day-to-day test-pilot-like life. Many chapters in this book can be directly attributed to the lab rats' hard work and experiences. Obviously they have earned the respect of tens of thousands of listeners of my radio show. I am the lab rats' biggest fan and supporter. They want for nothing except maybe a new owner who does not have a radio show or an overworked imagination and insatiable desire to learn more about horses.

One of the main reasons I chose to write "Box Stall Diaries" was to make it clear to you that caring for a horse properly is complicated by many factors. Age, sex, and breed play a factor in everything. Not to mention geographical location and the horse's activity level and overall health. When you ask for advice or guidance about anything to do with your horses, always be specific about what the situation is. Make sure the information you receive is appropriate to your individual horse and issues. A lot of what you read and hear about horses is presented as the definitive answer or rule to go by. Remember that versatility and individualism play a big part in your horse's life. Yes, there are some concrete rules and philosophies, I'll grant you that. Here is a list of day-to-day situations and health issues that should be considered on a horse-to-horse basis. Remember breed, sex, activity level and general health can change these. Out of convenience, horse owners, veterinarians and farriers know that they are a big deal but sometimes disregard these.

1. Dental Exams: Young and senior horses require more attention to their teeth: younger horses due to their fast development; older horses due to their decline in health.

2. Foot Care: Horses' hooves grow at different speeds. Also depending on whether the feet are dry or wet, thin or thick hoof walls. Does your horse just stay at home or does it travel and work?

3. Vaccinations/Immunizations: Where does the horse live? Where does the horse travel? Exactly what potential diseases and viruses do you need to protect your horse from?

4. Diet/Nutrition: How much hay? How much grass? How much grain? How much pellets? Any supplements? This is the essence of horse-by-horse thinking.

5. Weather Conditions: Horses handle heat and cold as differently as we do ourselves. Every horse seems to have a favorite time of year where climate and temperature are concerned.

6. Training Techniques: A lot of trainers and horse owners do not take each individual horse's temperament and personality into account when schooling a horse. I see this one just about daily.

7. Saddles, Tacks & Bits: It is very easy to assume one size or style fits all. Boy, if horses could talk. We would all hear about this one!!

8. Activities, Jobs to Perform: Not every horse can run barrels and poles, hit the brakes when a rope is dallied, jump a five-foot fence, win a race, or go up and down a rocky trail.

9. Horse-Human Relationships: Some horses either from a past incident or experience, prefer a man or prefer a woman. Actually now that I think about it, some horses don't care for any of us!!

10. <u>Being Sold/Traded:</u> New home, everyone is a total stranger. Nothing sounds or tastes like home. (Enough said. You understand this one!!)

I believe these ten can give you a pretty decent picture into just how easy it is to lose track of each horse's individual identity. Let me give you one more quick example – Does everyone in your family or your friends require their haircut on the same day? Do they all require the same food and nutrition at the same time? Amounts of sleep? Oh, this next one you will love. There are over 200 flavors of ice cream. My television has 180 channels. Varity, selection and an open mind. Here it comes. It's appropriate and a perfect place to say it. Are you ready??? Remember we are not talking about your time in this book. That's it. You guessed it. We are talking about "Horse Time"!!

In closing out this chapter, I would like to say thank you to my lab rats for teaching me about each horse's individualism and of course "Horse Time" as well. From my lab rats to you, always keep an open mind and realize every minute spent with your horses is an opportunity to learn.

Chapter Fourteen

"Halter Broke"

"Halter Broke" - what a misunderstood and overstated term. Halter breaking a horse is relatively easy. Basically, time and patience are the key factors involved. The younger an animal is, the easier it is to do. The reason is simple. No bad habits or mistrust to overcome. It's like the military; they like young minds that are fresh and easily molded. It's pretty much the same with horses.

Just about all training or behavioral problems in adult horses can be linked to their not being halter broken properly. Hard as it may be to believe, halter breaking a horse is more than just teaching it to lead. Halter breaking is a combination of discipline, training, overcoming fears, and love combined with respect. You don't want to break a horse's spirit, just channel it in the right path.

I will now describe in detail exactly what a halter-broken horse is. This list is probably going to make you think I'm crazy or I'm overstepping the bounds of everyday reality. Remember 95% of adult horse training or behavioral problems are due to their not being completely halter broken. The following is a list of what a horse should or should not do when halter broken.

1. Stands tied quietly for any length of time.

2. Accepts and does not walk away when approached with halter and lead rope.

3. Accepts shots and wormers with no acting up or trouble.

4. Stands calmly for trimming, shoeing, picking out hooves, etc.

5. Allows ears, nose, and body to be groomed and clipped.

6. Fly sprays and bathing are of no issue or concern to the horse.

7. Loads and unloads in a horse trailer just fine.

8. Does not walk all over you when feeding.

9. Respects your space while you are walking the horse.

10. Stands still for tacking-up.

11. Has no problem backing up with just a simple command.

12. Stands at your side without walking off or turning around pulling on the lead rope.

13. Stands while horse blanket is being put on or taken off.

14. Does not stop to eat grass, while being lead.

15. Does not try to nibble or bite you.

16. While eating, the horse allows you to separate it from food without a fight.

17. Steps away to the right or left (side pass) with just a slight push.

18. Is not startled by plastic or paper bags.

19. Is not prone to kicking when tail is brushed or braided.

20. Does not try to rush through an open gate.

Well, am I crazy? No. Actually think carefully about your own horses, friends' horses, or even just other boarders in the barn. You'll notice how these 20 items can really manifest themselves into bigger problems down the road. Now for the fun part. How many of those 20 behaviors does your horse display?

- 20 = Your horse is 100% halter broken. You're very lucky!!!
- 17 to 19 = Nobody's perfect. Your horse is just great!
- 13 to 16 = A little more schooling is needed.
- 10 to 12 = I recommend you go back to the basics.
- 7 to 9 = This is a good start; yet do not settle for this.

- 4 to 6 = I bet you know some really good curse words!
- 3 or less = Please reread this chapter.

Also depending on what you do with your horses, you may require even more abilities from them. Driving horses, show horses, ranch horses…every type of activity may need some sort of special training. Yet this list of 20 is a good foundation for any horse owner.

"Halter Broke", what a term. Now that you have read this chapter you may really see a lot of fundamental problems in horses that you thought were just fine. And at no age is it ever too late to try and improve on halter skills. Respect, trust and discipline are the building blocks to any good horse. Whoever said, "You can't teach an old dog new tricks" obviously wasn't a horse owner.

I can't give you a statistic or an average. What I can tell you is at least a couple of times a month, a letter or phone call comes to the radio show. Maybe someone comes up at a feed store or clinic, starts talking about a horse being killed – broken neck or choking due to a halter being left on a horse. Halters are probably one of the most important items to horse owners. Yet they need to be used in the proper manner.

Turning haltered horses out in pastures, paddocks, or even box stalls, can be dangerous. Halters can get hooked on anything: nails, branches, fence posts, gates, trailer hooks, bucket holders, hay and grain feeders. Once I saw a horse in a box stall that had died do to his leg being caught up in the halter. He was rolling or lying down and somehow put his foot through the side strap on the halter. These days' nylon halters are very, very strong. Good quality leather lasts a lifetime as well.

Halters are built to withstand all sorts of horse panic attacks and blow-ups. Also almost all manufacturers' halters are much stronger than any horse's neck. After you have left a horse unattended to roam around a pasture or in a box stall, how many times have you found a scrape, cut, or abrasion for no good reason.

Horses by nature are curious animals, always walking around, poking their heads in this place or that place. Don't forget how accident-prone horses are. Some companies are realizing that people leave halters on horses when they shouldn't. These companies are developing breakaway halters. Some are on the market now. Unfortunately the public has not yet bought very many.

Look, I really do understand that some horses are just about impossible to catch without a halter on. This is actually a training problem. The next time you are done with your horse and get ready to turn it loose in that pasture, or to close the door on that box stall, take a moment to remember this chapter and consider the possible outcome. I talk about this a lot to horse owners. The last thing I say to them really takes hold. "It's always easier to catch the horse than to dig a big hole and bury it."

I've discussed in detail the eventual goal you may wish to achieve and safety issues to always keep in mind. Now it is time to address how to start this process. First the basic tools needed: A halter that is "proper fitting" – proper fitting means not so loose as to move freely about the head and not so tight as to be binding on the head. The lead rope should be in good to excellent condition. No loose threads or weak snaps. Absolutely no knots in any of the lead rope. The lead rope should be no shorter than nine feet and no longer than 12 feet.

Now this next part is going to require a little extra explanation. Horses have a very powerful and sensitive sense of smell. Halter breaking can be very traumatic to a horse, especially wearing a halter for the first few times. The last thing you want is for this horse to have any excuse or reason to panic or just plain be uncooperative. So before you start your halter breaking training sessions wash the halter and lead rope in a washing machine. No soap, just hot water. Let them air dry completely. This will erase the scent and odor of any other horses. Or you can go to your local tack store and buy a new halter and lead rope. It does not matter if you buy a new halter or wash an old one. Well. Maybe. Depends on whose washer you use!!! ☺

The point here is to establish trust. Once you start using the halter and lead rope on your horse, they will become his. The feel, the scent, the odor will be familiar and non-threatening. Always use just that horse's tack. Also, obviously, don't use it on another horse as well. This also applies to brushes and combs. To a horse, smell is everything.

Ok, now let's talk about where to start this project. You may want to choose a round pen or arena. A 200-acre open pasture may not be your first choice. Why? Think about a horse running across that open field dragging a lead rope. Also don't look at me; I won't catch the horse for you.

Any forward movement while the horse is wearing a halter and lead rope needs to be acknowledged and the horse should be praised. If the horse starts to back up, go with it for several feet. Turn that rebellious motion into your command. Once the horse stops, praise the horse again. Make the horse think any movement is 100% under your control with the halter and lead rope as your guide.

Make frequent stops, also praising good behavior. Try keeping the horse still as well. Use a brush or grooming comb during this period. Touching your hands on the neck and shoulder areas is good too. Try rubbing your hands down the leg. Gently and slowly, no quick stuff or rough stuff.

The initial lesson should last 5 to 10 minutes, gradually working up to 20 to 25 minutes. After about a half hour, horses become bored and lose patience easily. Always end on a good note. This means make sure the horse is behaving and you are in command. This is the most fundamental key to any horse's training. The last memory, the last act of behavior must be positive and under your control!!

Now, hopefully, you remember a couple of pages in the front of this book. That's right, horse time. BIG-BIG Horse Time!! No rules on how long this should take or when to teach a horse. Halter breaking is the essence of horse time. There will be battles won by the horse. Also, never lose sight of the fact you are in horse time. You need to end on a good note. Also be patient and kind. Heavy hearted is ok. Heavy handed is not. You can lose a battle or two. Just remember you must win the war. This is a war of love, guidance and showing an animal how rewarding a friend you can be. Come to think of it, you know, a good pair of leather gloves might come in handy as well!! Rope burns can really make a sunny day seem cloudy.

Chapter Fifteen

"Round Pen or Lunging"

One of the most important and simplest of tasks for a horse owner to do is lunging a horse on the end of a 25-foot rope. Lunging a horse is not just the only way to train a horse; it's also by far the best way. Everything comes easy to a horse: where the water is, where the hay is, the sound of a rattling feed bucket, the noise of your truck and horse trailer. Horses learn very quickly what and where everything is. Inadvertent learning plays a big part in this process.

Working a horse in a round pen is key to just about everything. Teaching a horse to walk, trot, back up, and stop, and especially to respect you as the leader, is all best accomplished in a round pen.

Starting with the pen itself, you want to have at least 50 to 60 feet diameter, a nice big circle. Most lunge ropes are 25 feet long, so standing in a circle the horse will travel a 50 to 60 foot diameter path.

A smaller circle is tough on a horse. The best way to demonstrate this is to have you do a little experiment for yourself. Put a chair in the middle of a room. Walk a tight circle around the chair, then progressively start walking a larger and larger circle around the chair; notice the larger the circle the more balanced you are and the easier it is for you to walk. It's the same for a horse. About a 50 or 60 foot diameter is a pretty comfortable circle for a horse to move around in.

The type of fencing you use to make the round pen is up to you. You want something tall enough to hold a horse in, something that won't catch or hang up a foot. And a nice big gate. A tractor is handy to work up the ground with.

Let's talk about the ground. Sandy loam or soft tilled-up dirt. You want the soil to be loose and soft. Why? First, comfort and protection for the horse's joints and tendons. Next, to protect the horse's feet. Loose soil also tends to work the horse a little. This helps in muscle toning.

As far as how to train a horse or use a round pen correctly, there is an endless list. I can offer some general advice. Spend at least 20 minutes training, yet no longer than 30 to 45 minutes. Also you want to reverse direction, clockwise then counter-clockwise. If you decide you require a lunge whip, never actually make contact on the horse with the whip, simply snapping it should do. Voice commands should be used at all times. Pick a word or words for "go", "stop", "slow", "fast", and "reverse". Be consistent with your commands. And remember to praise and love your horse for good behavior. Don't forget to let the horse know and understand what is unacceptable as well.

Using a lunge rope and a round pen can be a horse owner's best tool for teaching and reinforcing good behavior. For a lot of us, time is tough to spare. Tacking up and riding everyday may not be possible, yet 20 to 30 minutes of round penning can keep a horse in shape and also remind the horse who the boss is. Good Luck!

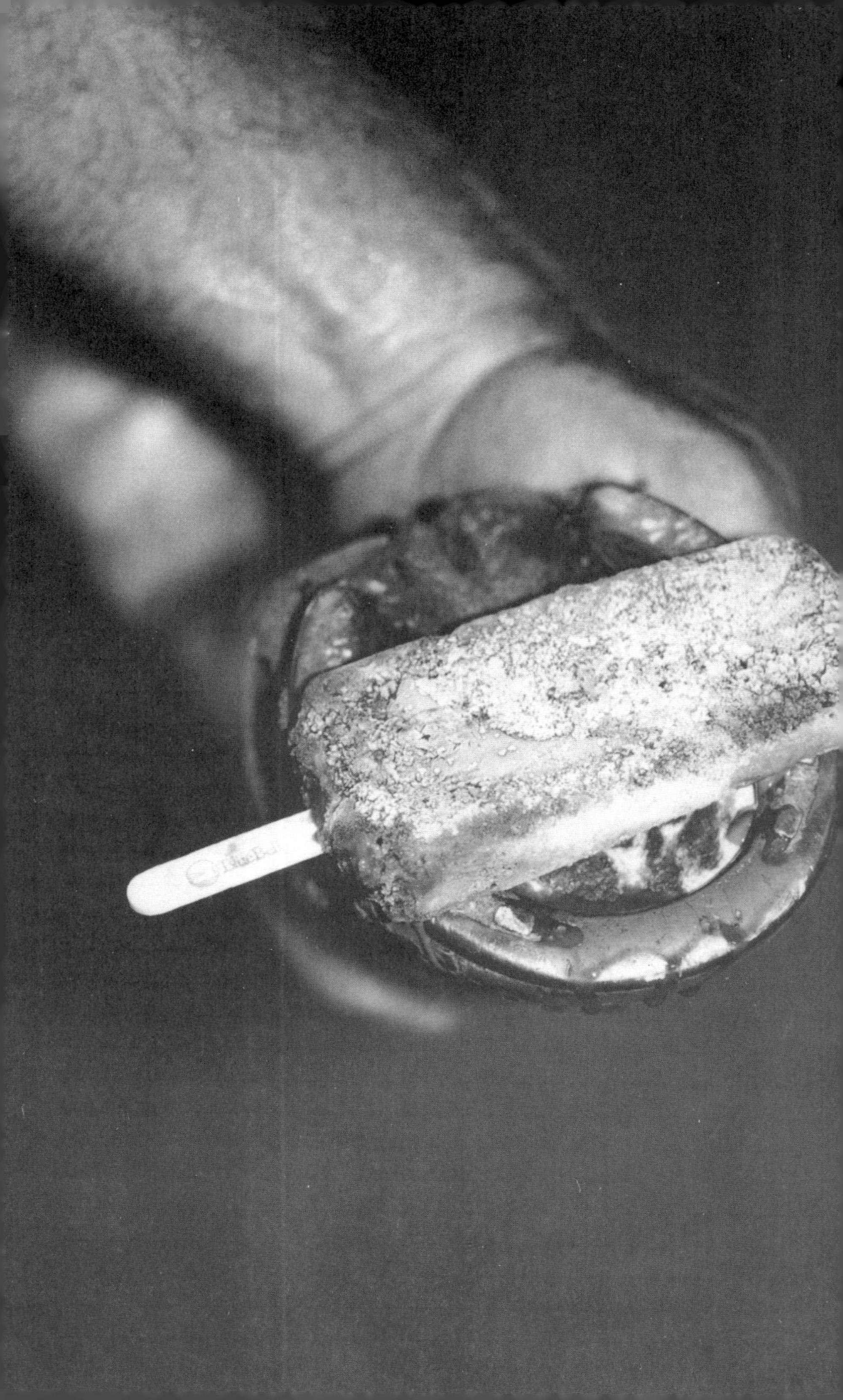

Chapter Sixteen

"Frozen Hooves"

You may think unless you live at the North Pole or unless your horse barn is a really large igloo, you really have no concern about your horse's feet freezing. Well, unfortunately you're mistaken, and here's why. First let's discuss the most important part of this situation. That would be temperature and weather conditions. Thirty-two degrees Fahrenheit is obviously the freezing point. Snow, rain, sleet, hail, and even mud can pose a problem at 32° F.

So now it's literally freezing outside. The second part would be the horse's hoof itself. Unless your horse is experiencing some sort of structural problem, the hoof is concave around the frog. This concave design is the reason every horse owner needs a hoof pick. The hoof is a natural shape for collecting and holding onto just about any type of soil or snow and ice it happens to step on.

Let's not forget, each time a horse takes a step, it causes whatever is in the hoof to compact, and compact, and did I mention compact? Compaction is what causes so many of the problems with a horse's foot. Now we add freezing cold air, with some sort of material on the ground to collect in the hoof and be compacted. Now you have a frozen foot.

Once it's there, it rarely removes itself. If it's snow, you can count on it getting larger. Also after awhile it starts to ball up and it becomes almost impossible for a horse to walk around comfortably. Not only is it hard for a horse to walk, a frozen foot also lowers the ability for the horse to stay warm. Constant exposure of the foot does expose the horse to other problems, such as stress cracks, or sole and frog bruising which can lead to abscess and infection.

It's vital any time of the year that you maintain your horse's hooves. In winter keeping the feet picked out and, more importantly, thawed out is key to the overall health of the horse. I think I've covered the reason why on this matter. Let me now explain what can be done to keep the horse's feet as clean as possible.

There is only one sure way to keep them clean and unfrozen. Keep the horse in a clean, dry box stall. However, that's not always an option. One very good option is a special pad called an anti-snow balling pad. It is a hard plastic pad designed to push and remove material out of the horse's hoof. These pads are used in conjunction with horseshoes. Unless the shoes have caulks on them or borium or Drill Tek, shoes can be slippery on ice and snow.

Another way to help prevent the collection of material in the foot is to apply some sort of slippery thick product to act as a barrier in preventing snow, ice and mud from sticking to the bottom of the foot.

Here is a list of commonly used products that have proven themselves to be useful in keeping a foot clean.

1. Vaseline
2. K-Y jelly
3. Pam cooking spray
4. Crisco cooking lard
5. Any type of hoof moisturizing paste
6. Hand lotion

This is very important, folks. Never, ever use any type of lube, grease or motor oil; no petroleum based products at all. If you think it would cause damage to your skin, then it is also harmful to the horse's foot. Just like our skin, the horse's hoof is sensitive.

In winter you need to check your horse's hooves every day. It's best to clean and pick them out twice a day if you can. Remember compaction is what you need to try to avoid. I hope your winter goes well and you can keep your horse's hooves thawed out.

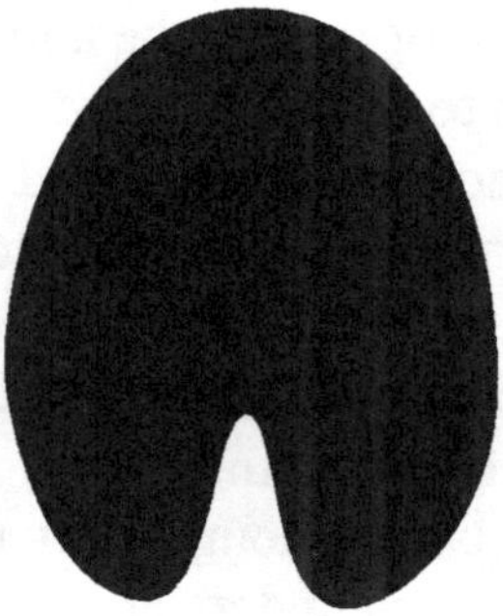

Chapter Seventeen

"Hoof Compaction"

Obviously it doesn't make much sense to lift up a horse's hoof, pick it out, and then put the hoof back down in the dirt. Why clean it out if it fills up with dirt right away? Well, let's talk about the horse's hoof, how it works and why hoof picking is extremely vital.

First, the outside of the foot is called the horn. It's the area where nails are placed. This area is also painted and shined on show horses (a lot of work to do this).

The bottom of the foot has five distinct areas. There's the sole – this should be concave. The outer edge, the area that is trimmed and where the shoe is applied, is called the white line or laminae. That big triangle-shaped piece is the frog. At the back of the frog are the bulbs of the heel. Following the hood edge around to the back of the frog, next to the heel bulbs, are two distinct pieces of hoof angling in towards the frog. These are called bars.

Now let's talk about compaction. Every time a horse lifts his foot and sets it down, the hoof expands and contracts, just like your feet. Every time you take a step, your foot expands to absorb your body weight. The horse's foot does the same thing. When it fills up with dirt, manure or debris, the ability of the foot to function properly starts to drop.

Constant compaction of dirt placed on a horse's foot does several things. It dries the foot out, starts to break down the frog with pressure, and creates loss of traction. Compacted dirt also places more stress on the horse's foot bones and lower leg bones than they are designed to withstand.

Racing, roping, barrel racing, showing. Most riders who participate in these activities understand and are aware of compaction and the importance of picking a horse's hooves out. It doesn't matter who you are or what you do with your horse, picking out the feet is very important. I always tell folks you're not doing these things for your benefit; you're doing them for the horse's benefit. The more time you spend on the "little to-do's "the greater the chance you'll spot any minor problems before they become major ones.

It's really a pretty simple philosophy. You look in the bathroom mirror everyday. You spot every little change in your face, however slight. Well, the closer you get to your horse the more you become familiar with your horse's body. The easier it is to spot cuts, scrapes, sore joints, swelling, eye trouble or anything out of place.

Over time, you're gonna find that the horse's feet have much more to do with their overall health than you can imagine. Picking them out is the simplest, quickest and most expedient way of starting on the road to healthy feet.

When I was a kid, there was this old cowboy; he had the most disgusting stories about horses. This man could really ride, yet he had no social skills at all. I used to listen to everything he said. I still can hear his gravelly voice as I was picking out a horse's hooves.

"It's just as easy to pick a horse's feet as your nose." Just for the record, folks, I'm always picking out horse's feet. Nothing, ---- I repeat, ---- nothing else! ☺

Chapter Eighteen

"How Dry is Dry?"

Probably one of the biggest and most troublesome aspects of taking care of a horse is the feet. It doesn't take long for a horse's hooves to start to get hard and brittle due to the lack of moisture.

First let's discuss how much water is actually in the foot. Each part of the foot has a different cell structure and moisture content. The outside of the hoof is about 8% to 12% water in its makeup. The sole is about 15% to 20% water and frog is 30% to 40%.

It's awfully easy to dry a foot out. Water is extremely important to the overall health of the horse's foot. Buying hoof treatments can help, yet only on to very small extent. Once a foot is dried out, store-bought hoof care products really are facing an uphill battle. Hoof care products are great for maintaining an already healthy foot, but expecting any product smeared on once a day to remoisturize a foot is foolhardy.

Really, the only way to bring back what is gone is to soak the horse's hooves in clean cool water. The water only has to be one to two inches deep. Remember we are trying to get moisture into the sole and frog. Having the foot completely submerged in water can be risky. Yes, it will help moisturize a foot faster, yet getting the coronary band too soft is bad.

The best way to tell if your horse's feet are becoming too dry is to perform a little test. Try taking your thumbnail and scraping away at the sole and frog. If you can peel and scrape the frog and sole, you horse's hooves are probably all right.

If the sole and frog are hard and stiff, you need to think about softening up the feet. The more pliable and soft a horse's feet are, the fewer cracks and chips will occur. Also remember that a horse's foot is supposed to absorb body weight as well. The healthier the foot, the better the foot will function. It will also hold shoes longer and grow faster. Just about everything will be more positive.

Folks, you can spend a lot of money on hoof treatments and feed supplements. And lose a lot of horseshoes on dry feet. Try to use this chapter to adjust the way you think about your horse's feet.

I personally think we figure a horse's foot is like ours. We function fine with foot trouble, yet horses can't. The comfort and condition of the foot can even cause a horse to stop eating. A horse's body weight is only supported on four little feet. Just four points of support. It's about a 60/40 split: 60% on the front feet, 40% on the back feet. That's one reason why horses suffer from more front-end problems than rear-end problems.

Shoes and trimming really can't do much for a horse's foot, unless the foot itself is in good shape. Obviously, good shape means plenty of moisture and a balanced diet. A balanced diet is also what seems to escape the foot. Try to keep your horse's daily intake of food on a balanced level. Not just protein and carbohydrates, but also plenty of vitamins and minerals.

In closing, there's an old barn saying "no hoof – no horse." It really does apply and is extremely accurate.

Chapter Nineteen

"Proper Hoof Trimming"

If you spend any amount of time around a horse barn, sooner or later someone speaks up with great conviction and authority when stating, "No Foot – No Horse." Unfortunately my experience has been this person is usually the one person in the barn that just irritates everyone. So even though "no foot – no horse" is extremely accurate and a convincing statement, rarely is it understood or taken seriously.

There are some very basic criteria for the horse's hoof. First, the horse's hoof is not like our fingernails. That's just plain ridiculous. You lose a toenail; you can go about your life just fine. You just wait for it to grow back. A horse loses a hoof; that horse will probably die. The hoof of a horse has a very important role in the day-to-day existence of the horse. A horse's appetite, energy level, and personality all can be affected by very minor foot problems. What I intend to do is not teach you how to trim and shoe your horses. Yet this chapter will teach you how to know if your horse's hooves are being trimmed and shod correctly. There are the angles of the foot, the length of the toes, and the ability of the horse to stand on a flat, straight foot.

Angles refer to the amount of elevation in the relationship of the toe of the foot and the heel of the foot. Arthritis in joints, bowed tendons, tripping or stumbling, just about all of the injuries or bone disorders in a horse's lower body can be created or prevented by the angle of the foot. What is more important is to realize that the way the foot is trimmed is usually the cause of the problem rather than the shoe itself.

Obviously some of the more popular horse activities under the best conditions still take their toll on a horse's body. Barrel racing, horse racing, jumping, and roping tend to be the most damaging. Let me qualify this by saying they are also very popular and growing horse activities.

What I am now going to show you are the ideal angles and toe lengths. These apply to all horse breeds, large to small, even donkeys, mules, and zebras. The only exceptions to these rules are special physical activities that a horse is required to perform which cannot be accomplished without making adjustments to the natural, most ideal size, shape and angle of the foot. Think about that. That's why performance horses have so much trouble. A horse is required to do more work with an uncomfortable foot position. Would you want to move a refrigerator out of the house wearing a pair of swim fins? I realize that sounds a little odd; yet think about that.

Front feet should be the same length and angle. Back feet should be equal as well. Here is the tricky part; front and back feet do not match up. Generally front feet are longer toed than the back feet are. Think of it this way. On your truck you can have two different tire sizes and air pressure levels. Front tires versus back tires. Yet all mounted and balanced properly, the truck rolls down the road just fine. There are devices called hoof gauges to measure the angle of a horse's hoof. They cost about $25.00 to $60.00. Why a lot of people trim horses without using one is beyond me. The hoof gauge cannot tell you what the perfect angle is for your horse. It just makes sure both feet are set at the same angle. It's a type of ruler; you can saw two boards without a measuring tape. If you cut wood a lot you can come close after a while. Yet ever notice how much a really good carpenter measures each board before cutting it? This guarantees two boards to be as exact and even as humanly possible.

The difference of 1/16 of an inch between the right and left foot can cause a horse to stumble or just miss coming into the money at any competition. Now do you really feel at ease when the person trimming your horse's hooves says, "Don't worry, I've been doing this so long I can just eyeball it"!!!

Trace a ruler and measure straight down the front of your horse's hooves. Start at the coronary band and measure downward to the tip of the toe. Unfortunately a lot of you who do this are going to be surprised and upset. Just imagine how your horse feels walking on uneven feet. It would be like you walking in a work boot on one foot and a house slipper on the other foot. Measure right and left fronts to line up and match. Then right and left rears to line up and match. Remember keep fronts and backs separate. It's apples and oranges. You want twin apples and twin oranges.

Next stand directly to the side of the horse and lift the foot off the ground. Hold it low and look straight down at it, and compare the outside to inside of the bottom of the foot to see if it is trimmed flat and true. This is like looking at the bottom of your boot at the heel to see if one side or another is worn down. This makes you stand and walk at either an inward or outward slant. If the horse's foot is longer on one side than another, this means the foot may be trimmed crookedly. This also can slow a horse down or make the horse ride rough. Now that you understand how important angles and foot lengths are, let me show you how you, the horse owner, can check your horses at a glance to see just how well your horse's hooves are being taken care of. No gauges. Those are supposed to be used by the professional. Notice I said supposed to be used.

Stand your horse on a flat surface, concrete or asphalt works best - examples are patios, driveways, parking lots, alleyways of barns. Make sure the horse is standing at attention with all four legs under him evenly, just like you would for a halter class show or photograph.

Next, step away from the horse and to the side about 15 feet or so. The pastern bone and hoof should be at one straight angle. This applies to both fronts and backs. Remember not all four angles are the same. We want twin apples and twin oranges!!

Unfortunately again a lot of you are going to be surprised and upset with what you find.

1. A steep angled pastern and a long, low angled foot means the horse needs shorter toes and longer heels. Less trimming on the heels.

2. A low angled pastern and a high short angled foot means the horse needs shorter heels and longer toes. Less trimming of the toes.

3. If you see a nice even straight angle down the pastern bone, down the hoof to the ground, this means your horse is at natural healthy angles. Trimming is probably pretty good.

Now about the horse standing on a flat straight foot. Pick up the horse's foot; hold it low and look over it and notice the shape. Compare the outside edge to the inside edge; pretend the horse's hoof has a clock on the bottom.

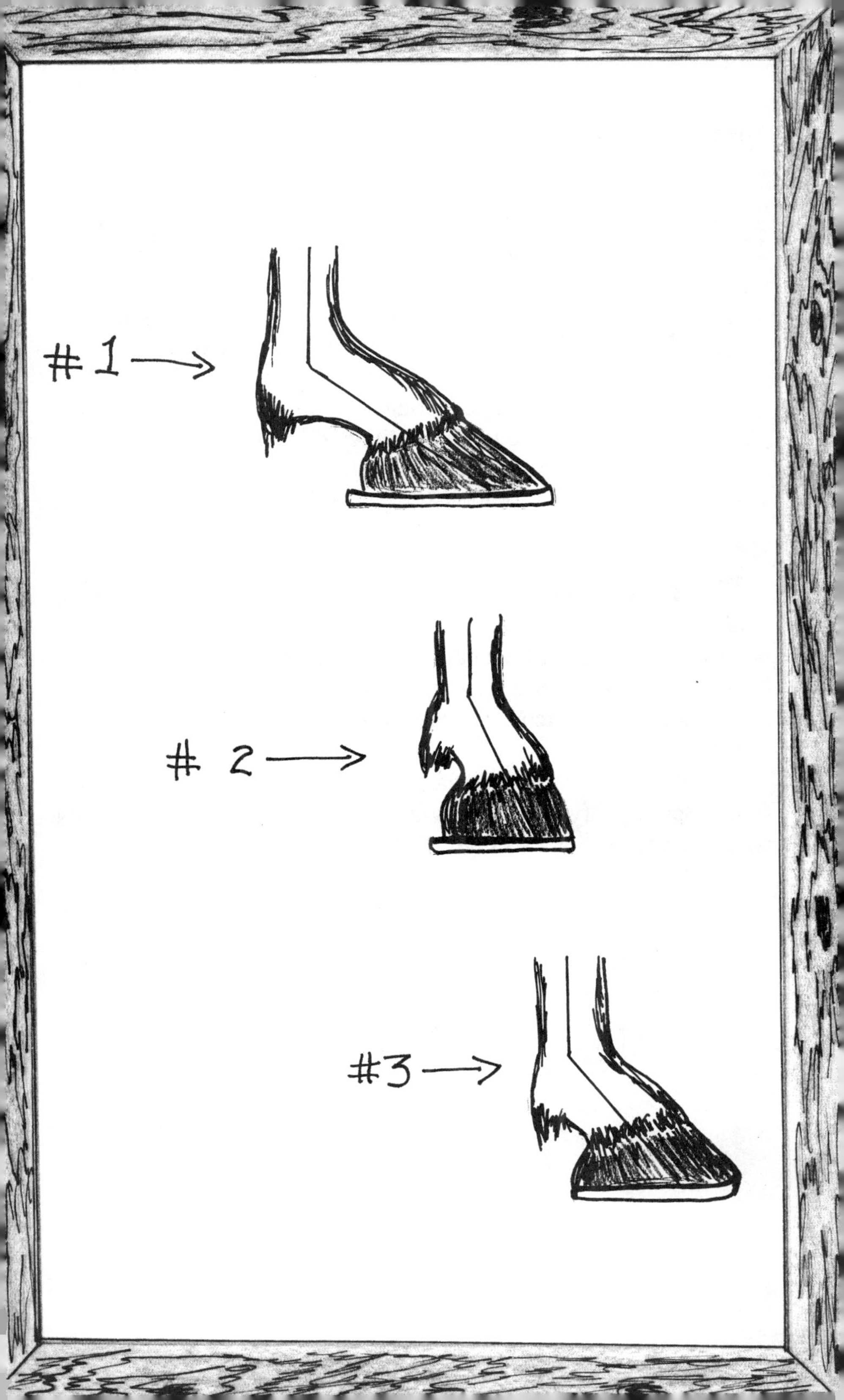

#1
#2
#3

The toe is twelve; the back of the frog is six. You want to compare 12 to 6 and 6 to 12. The foot should be flat, straight and true.

1. If the inside is longer than the outside, this means the horse is leaning outwards on its foot.

2. If the outside is longer than the inside, this means the horse is leaning inwards on its foot.

3. If the foot looks flat, then the horse's foot is probably being trimmed correctly.

If after doing all of this, you find some feet are at different angles and other feet are trimmed incorrectly, then you need to have a discussion with the person you have entrusted with your horse's hoof care needs. Let me say that ignorance or just not knowing does not excuse your responsibilities as the owner of the horse.

It's your horse, your money. You have the final say and final responsibility to do the right thing. Obviously you do; that's why you are reading this book!!!! ☺

The diagram on the next page displays how much difference can be achieved by properly trimming the hoof.

- Notice the angle in the bottom hoof how much more natural it looks.

- This was achieved by trimming more toe and less of the heel.

- Notice the top hoof how the sketch displays the amount trimmed off the hoof.

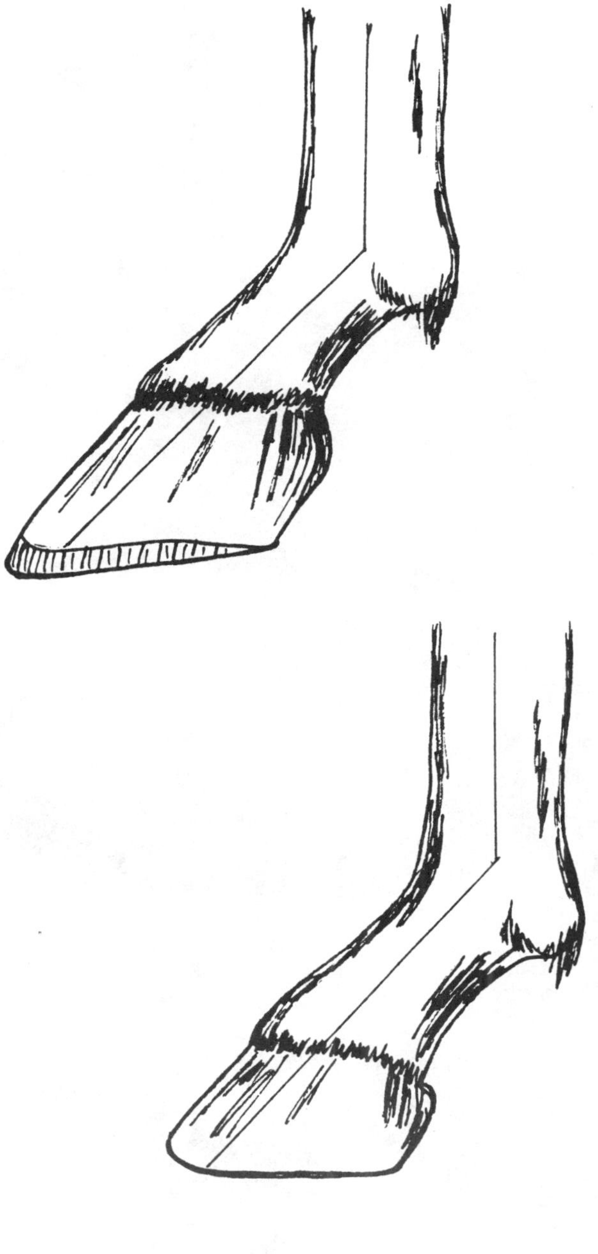

AL 01872651C

Chapter Twenty

"Proper Shoeing"

I hope the last chapter on trimming was easily understood. Now we are moving on to the other important part of the horse's hoof. That's right, shoeing the horse. No need for shoes or an anvil or a hot forge here! Not even a sore back in sight.

The biggest misconception about horseshoes is how and why the shoe stays or comes off the foot. The most popular theory is the higher up the foot you nail, the more solid and sturdy the hoof and shoe will be. That is just not the case at all.

The relationship of the shoe to the foot is key to this dilemma. There cannot exist any voids or unevenness between the shoe and the hoof. Even as hard and tough as a horseshoe is, the shear body weight of the horse makes a shoe expand and contract. Also there is flexing involved as well. Oh sure, I'll be the first to admit this movement is on a small, almost minute, level. Compound this activity over 24 hours, seven days a week. It's easy for a shoe to work its way loose. Once a shoe is loose, it is not long till it's gone!!

There are very simple common sense reasons that explain how and why a foot can hold a shoe. Let's think about this. You

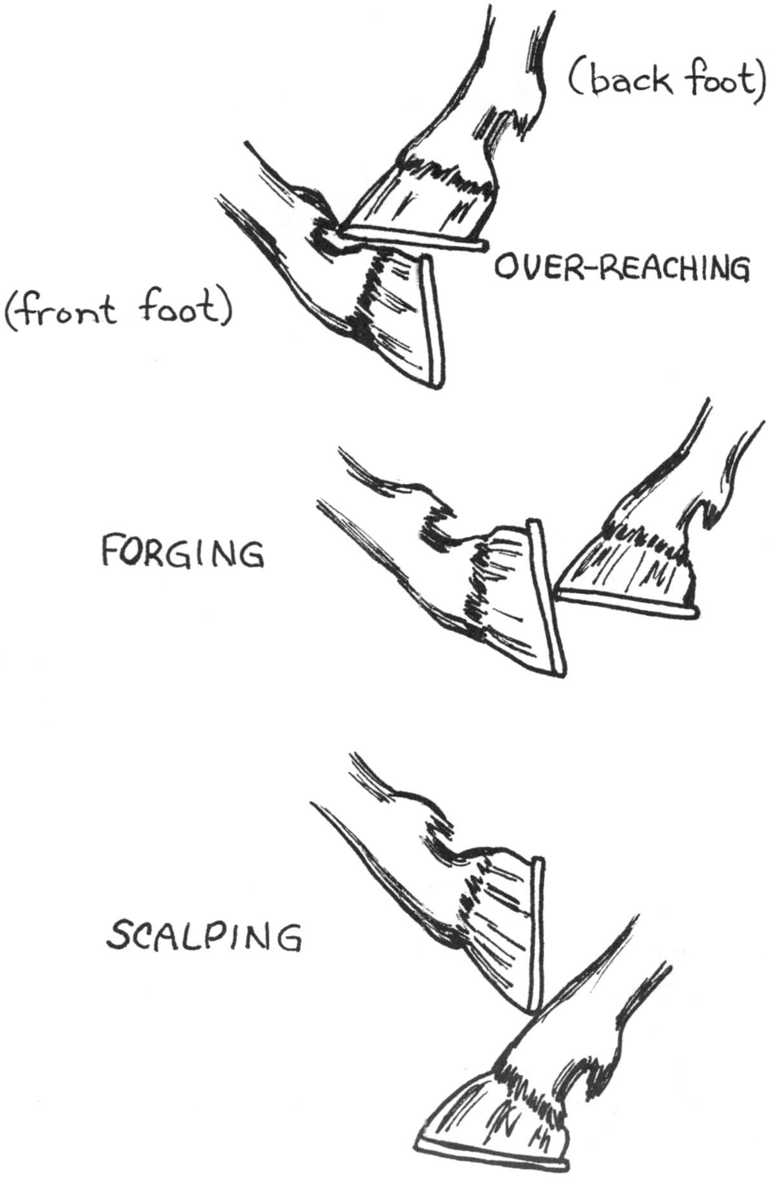
(back foot)
OVER-REACHING
(front foot)
FORGING
SCALPING

take a nice healthy foot and hammer small slivers of metal through it to secure larger pieces of metal to the foot. Boy, that sounds so inviting; makes you just want to run out and do it now.

The nails should not be any higher up the hoof from the shoe than ½". The goal is to trim out any nail holes that are not being used. The fewer nails, less dirt and bacteria get in. Also, the fewer the holes the stronger the actual foot is. Let's not forget how easy a horse's hoof dries out in the summer. Holes cause cracks, chips and dryness.

If the relationship of the hoof and shoe is correct, there will be no movement. Less movement, the less stress there is on the horseshoe nails. Put all this together, the shoe stays on the foot. Like just about everything with a horse, there are a lot of exceptions to shoes staying on.

1. If angles and the lengths are wrong, the horse may make contact with front and back shoes.

2. If the shoes are too big for the foot, they will not stay on the hoof very well.

3. Thick, deep mud can suck a shoe off.

4. The horse can do something stupid like back out of a trailer, catching the shoe and tearing it off or pawing at a fence or gate and pulling it off. (Obviously I am not talking about your horse; it's someone you don't even know. This statement applies to everyone who reads this book!!!) ☺

This list goes on and on and on and on. I want to be very clear right now. There are many medical and physiological reasons that can take your horse's feet into unbelievable areas: type of shoes, trimming, placement of nails, or even possibly glued on shoes. This chapter is written to help you gain the ability to feel more confident dealing with your horse's hooves, whether you ride and own performance horses, trail riding horses, show horses or gaited horses who must move in a specific manner. All these variables can dictate the way a horse is shod. There are several hundred types of shoes and nails and different metal alloys as well. When you combine all this into one big bucket, that should show you how important your horse's feet are to the overall health of the horse. Confidence and peace of mind are very crucial when referring to the person working on your horse's hooves.

A final thought about trimming and shoeing: experience and knowledge must be combined with the right tools for the right job. Should this be the basis for choosing someone to work on your horse's hooves? Or should it be like shopping for groceries – who has the cheapest prices? Price does not always denote or guarantee quality. I'll grant you that. Yet, is it a step in the right direction? Remember I said early it's your horse and your money!!!

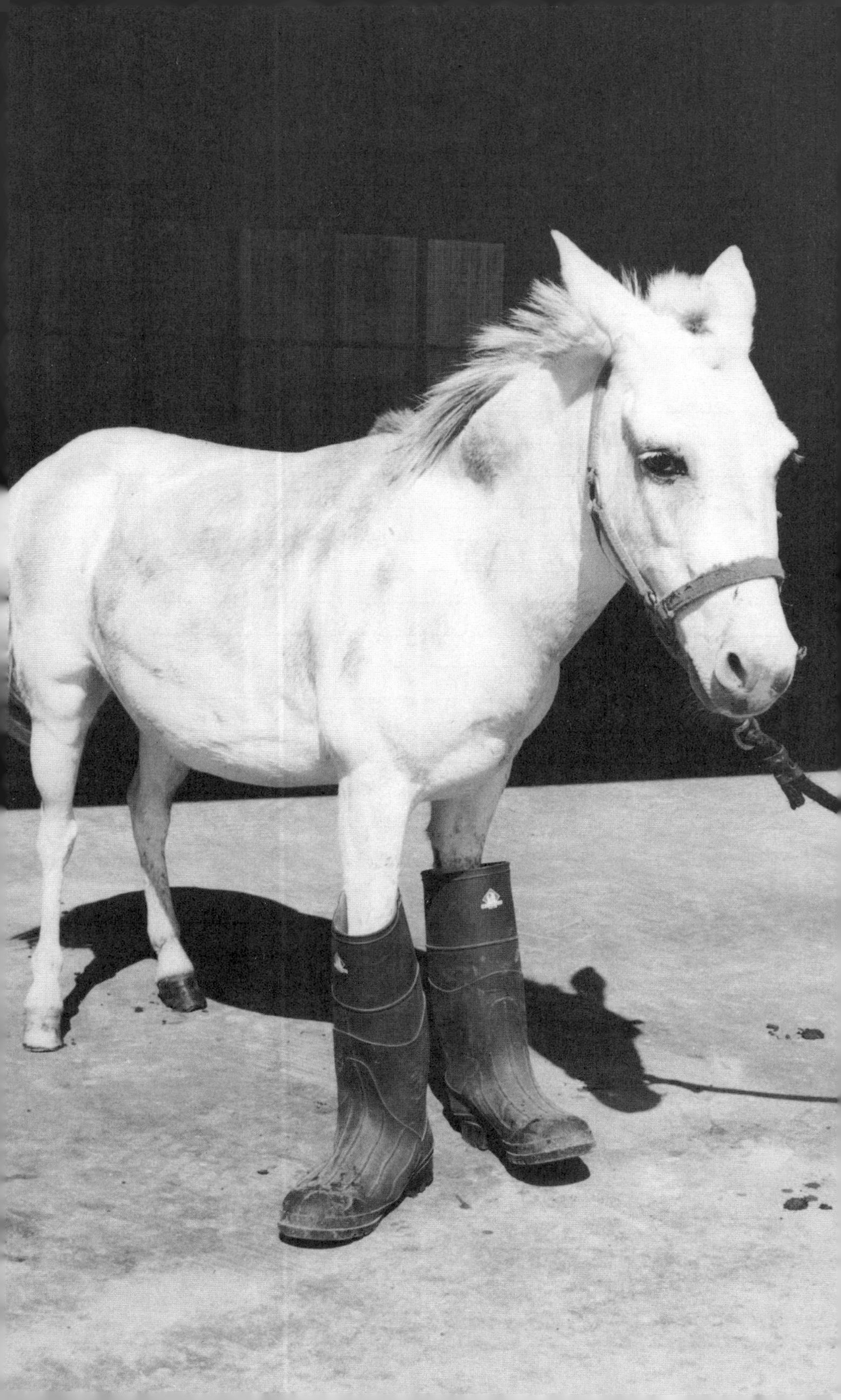

Chapter Twenty-One

"Winter Traction"

Just because winter is coming doesn't mean we don't want to ride!! Yet a horse slipping and sliding on snow and ice really doesn't sound good. A regular horseshoe, whether it's steel or aluminum, won't offer any help in winter conditions. Actually, I've found ordinary horseshoes provide less traction on snow and ice than a horse just being barefoot, just like riding a horse in the 4th of July parade down Main Street, not much traction when you need it.

Is there winter traction? Yes. Is it easy? Yes. Can you afford it? Yes again. Actually you have a choice of three types of traction applications on your horse's shoes. All you need to do is understand what they are and how they work.

Let's start with the basic premise that you are going to need traction on ice and snow-covered roads and trails. Also you want simplicity and economy as well. Here is a list of the most common and most widely used methods of gaining winter traction for a horse.

1. "Caulks": (pronounced corks) You hunter jumpers will know these well. Caulks are small metal cleats either screwed or driven into the heels of horseshoes. There are many sizes and styles available. The cost is usually under $1.00 each. The horseshoer's labor for this procedure is minimal labor. Very good traction, yet they do wear out with use.

2. "Borium": A metal alloy that is applied to the shoes with heat. Basically a clump of it is melted on the shoe. Not hard to apply, it's readily available. Great traction yet can chip or break off with use.

3. "Drill Tek": Same type of situation as borium. Harder to apply, the most costly of the three. Yet if applied properly will not – I repeat – will not chip, break or wear out. Fantastic traction in all terrains and weather conditions.

All three must be done by an excellent horseshoer. Applying anything to the bottom of a horse's shoe can change the way that horse moves. You can actually make the foot tender if the shoe is uneven due to traction devices not being applied correctly.

There are other devices on the market. Pull-on "snow boots", clamp on "horsey snow spikes" and there are specially designed horseshoe nails. All have their place, yet the three I listed offer the most traction and longest wear.

The other issue with winter traction is the sole and frog; how to keep snow and ice from building up on the bottom of a horse's foot. There is a special "anti-snow pad" you put on with a shoe. They actually work well. Yet most folks like a horse without shoes and pads. Shoes are ok; pads can be a little problem.

Believe it or not, spray cans of cooking oil work the best. The stuff you spray on a frying pan so eggs don't stick - great stuff for keeping snow and ice off the feet. Other things like baby oil, corn oil, vegetable oil, and Vaseline petroleum jelly, all these work. You just apply a good thick coating on the bottom of a clean and dry horse hoof. Try not to use automotive products; they tend to hurt a horse's foot.

Well, folks, I wish you a mild winter. Consult your local horse doctor and farrier on winter traction devices. Remember. Left to roam, horses migrate south to greener pastures in bad weather. A horse is a horse, not a snowmobile.

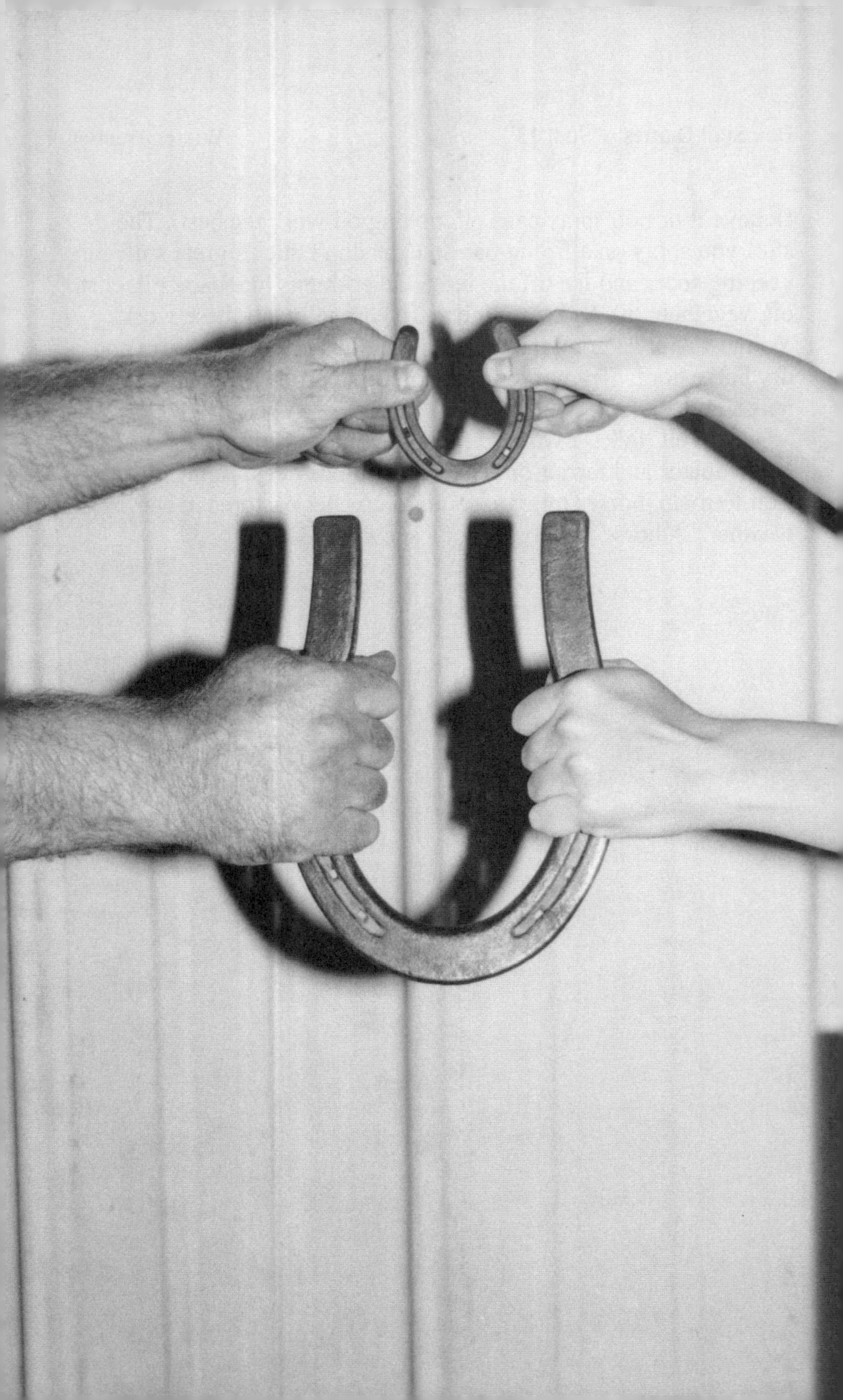

Chapter Twenty-Two

"The Horseshoe Switch"

Farriers and horseshoers can be a dime a dozen, but finding the right one for the job can be next to impossible. Although I must admit there are a couple of buckets full of nitwits and people who should never touch a horse, there are also folks who take their craft seriously.

I want to give you some food for thought: the average full-time shoer will probably have from 5 to 12 horses a day, 5 to 6 days a week. So that's about 30 to 72 horses a week or 120 to 288 a month or 1,400 to 3,000 a year. Odds are, something will go wrong. Nobody wants it to happen; yet it's going to occur. Nobody is perfect, especially someone working on an animal that can move at a moment's notice. It seems that switching farriers is common, especially when something goes wrong: a hot nail or wrong shoe, maybe the foot didn't get trimmed right.

Granted, you need the horse's feet perfect; you can't afford mistakes. A lot is riding on every show, each event and all the rodeos. You need to have trust and confidence in your horseshoer. Having good lines of communication and understanding is key to any successful horse operation. Yet we are a very unforgiving lot.

One mistake to some owners/trainers is one too many. So the hunt is on for a new person to trim and shoe. Sure, sometimes there is a person so incapable of doing a good job, they really need to find a new career. Yes, drop these folks real fast. You need quality work done on your horses, as consistent as humanly possible.

Ladies and gentlemen, mistakes are going to happen. The horse moves, it's a bad day, whatever. It happens. Try to work with your farrier. A lot of problems are just plain miscommunication. Sure, some egos do get in the way, along with attitude. Cooperation is the key in any area of horse work. Try to keep all of this in mind. The longer a farrier works on one horse, the better understanding he/she has of each horse's needs and not just the horse's needs but the rider's needs as well. Time and hands-on experience are hard to beat when it comes to a horse's hoof-care needs.

Well, I wanted to give you something to think about. Maybe just a small thought to ponder when considering switching farriers. Also on how to deal with them. In our business, as with most competitive things, egos and attitudes abound everywhere.

Try to look at the big picture and really sort things out. You may have noticed, I said nothing of horseshoeing fees. I could have said, "You get what you pay for," or "They're too darn high." Maybe even, "It's the best money you can spend caring for your horse!!!"

"The measure of a farrier is not his best work. It is how he deals with and fixes his mistakes."

LIBERATOR A/T

Chapter Twenty-Three

"Buying Truck Tires"

You're probably wondering what buying truck tires has to do with your horses; if you never put your horses in a trailer and haul them, absolutely nothing. If you do have a trailer and you do haul your horses, then this chapter can help you out. And save you lots of money.

When you buy a new truck, or you buy tires for a truck you've owned for awhile, very rarely does trailer weight or hauling hay and grain even enter the picture. But it should, especially if you consider the cost, of the horses, the cost of the trailer, and, most importantly, the value of your life and the lives of other drivers on the road.

Most vehicles have tires that are safe and sturdy enough for the unloaded vehicle only. Start stacking hay in a field, loading grain and sawdust at a feed store, hitching up your trailer, and then putting horses in it.

It doesn't take long to "max-out" your tires' carrying capacity. Every tire has a maximum gross carrying weight printed on the sidewall of the tire. Every tire has a load range code. It can be load range C, D or E. Load range refers to the amount of load a tire is rated to support and carry and not fall apart while rolling down the road.

When shopping for tires, explain to the tire store that you haul hay and grain, and that you haul a horse trailer. Know how much your trailer weighs, including the horse weight. All these weight factors need to be not only considered, but also absolutely included in buying tires. You may end up with too much tire when just driving around in an empty truck, but the best thing is, when loaded, you have what you need.

The easiest way to find out what your truck and trailer plus tack and feed weighs is to load up, hitch-up and drive to a truck stop with a scale. Usually a truck scale will cost only a few dollars. Truck scales are open 24 hours, seven days a week. After you know what your total outfit weighs, you can very easily go to a tire store and purchase what you need.

Try to buy tires that can be purchased or replaced within reason just about anywhere you may possibly travel. That way, when away from home, on the road, if you have a road hazard or problem, you can get the same tire you are rolling on.

Best of luck and check your tires.

Chapter Twenty-Four

"Water Tank Heaters"

Every year it seems we all spend just a little too much time and money on tank heaters. Electric, gas, wood – whatever the energy source – tank heaters keep our horses' water thawed out. Also tank heaters really get on our nerves. I'm sure a lot of you know this through past experiences. First let's talk about purchasing a tank heater. Tank heaters are available just about everywhere – feed stores, elevators, saddle shops, hardware stores and catalogs. The electrical 110-volt models really aren't very costly; prices range between $18.00 and $45.00.

There are heated five gallon plastic buckets that cost $25.00 to $60.00. You can also get natural gas or propane powered livestock tank heaters. These average out at $300.00 to $500.00. Wood or coal burning units are costing about $150.00 to $225.00. Following is a breakdown of all the different types of heaters and the aspects of each one. This list will help you decide which type and model best suits your winter weather water needs.

1. Floating Tank Heater
   Average Cost: $18 - $40
   Power Source: 110 volts

   Can keep 100 gallons of water semi-thawed in light to medium cold conditions (32° to 0°). Easy to use, very low maintenance. Downside: horses love to play with them as they float around on top of the water. They're safe in any tank (metal or plastic) with a heating element guard.

2. Submersible Tank Heater
   Average Cost: $22 - $50
   Power Source: 110 volts

   Can keep 100 to 150 gallons of water semi-thawed in light to medium cold conditions (32° to 0°). Easy to use. Tank must be kept clean since heater sits on the bottom of tank or mounts in drain plughole. Safe from horses since it's under water. Protect tank from elements by setting heater on bricks or using heating element cage.

3. Natural Gas-Propane Heater
   Average Cost: $300 - $500
   Power Source: Natural or propane gas

   Can keep up to 500 gallons of water semi-thawed in heavy cold conditions (32° to -20°). Large units are heavy. Mount on side of tanks. Must fill gasoline containers. Perfect for larger tanks out in the open.

4. Wood or Coal Heaters
   Average Cost: $150 - $220

Power Source: Burning wood or coal

Can heat a large tank or even keep a large hole open in a pond or stream in heavy old conditions (32° to -30°). Units are basically a heavy metal burning unit. Will not harm livestock, yet you have to check the fire three times a day. The colder the weather, the hotter and longer lasting you want the fire to be. The oldest of all tank heaters.

5. Plastic Bucket
   Average Cost: $25 - $60
   Power Source: 110 volts

   Will keep water semi-thawed in light conditions (32° to 20°). Works well in stalls. Outside conditions tough on buckets over time.

6. Submersible Bucket Heater
   Average Cost: $18 - $25
   Power Source: 110 volts

   This is a small heater that you stick in a bucket. Best for use in metal buckets. Great for short-term use.

7. Water Recyclers
   Average Cost: $35 - $75
   Power Source: 110 volts

   Unit does not heat; it circulates water and causes waves. Theory is to keep water moving so it won't freeze. For use in mild conditions at best (32° to 25°).

Each type of heater works well. The problem is folks tend to use them incorrectly with the wrong tank size and/or weather condition. Even the tank's location can make or break a heater's chances of working properly. Keeping a water tank out of the wind, preferably inside a barn or shed makes a big difference in its ability to keep water thawed. Also providing a wooden cover that covers half the open surface of the tank will keep water from freezing. Separating the tank from the ground with hay or straw creates an insulator to keep water warmer.

The most important thing to do with any and all electric heaters is to properly ground the tank and water. Electric tank heaters have a tendency to emit just a fraction of electricity into the water – not nearly enough to kill a horse, just a little tingling feeling. The little electrical charge on the horse's nose discourages the horse from drinking. And if for some strange reason the heater was to fail, a tank properly grounded would go a long way in protecting your horse's life. Most tank heaters come with a detailed diagram and instruction sheet on grounding a tank. Plastic, steel, aluminum, and even wood water tanks need to be grounded.

Installation and Grounding a Water Tank

Supplies Needed:

1 – 8ft long copper grounding rod
1 – grounding rod clamp
1 – 3/8" or ½" x 1 ½" nut, bolt, washers
1 – a spool of stainless steel electric fence wire (at least 20 feet)
1 – 2 to 3 pound cement or clay brick

Installation Instructions:

A) First place your tank in the location you want for the winter.

B) Drive 8 ft. grounding rod into the ground at least 7 feet deep right next to water tank. (Keep rod close to tank so horses won't get caught on it.)

C) On Metal Tanks: drill a hole through the side, right under the top lip of the tank.

D) Insert nut and bolt into hole in metal tank, attaching the stainless steel wire around bolt. Use a washer to secure wire contacting the tank wall.

E) For wood or plastic tanks: wrap 2 feet of wire tightly around brick. Put brick in bottom of tank.

F) Once you have the wire attached to the side of a metal tank or attached to a brick on the bottom of a wood or plastic tank, secure wire to grounding rod with the clamp.

G) Your tank is now grounded. Also, protect the heater's electrical cord from your horses by sliding PVC pipe over it, or by using lumber to hide the cord.

H) Try very hard not to use extension cords. Manufacturers of tank heaters discourage their use.

If you run into trouble when installing a heater or grounding system, consult your local feed store or contact the manufacturer of the tank heater. It's easier to ask questions than to explain how your horse was electrocuted.

Chapter Twenty-Five

"Trailer Traveling Tips"

Most horse people take traveling with a horse trailer for granted. Unfortunately, you never really know what you need until you really, really need it. Parked on the side of the highway is not the time to second-guess yourself. Being a little prepared goes a long way in preventing disasters. Also, it could mean the difference in making a show or rodeo at all, or just limping on home.

Let's just list things to think about and items to carry with you – ones that can really make a difference in a traveling situation.

1. Have a jack large enough to lift trailer tires off the ground with horses in the trailer. Do you really want to unload your horses on the shoulder of the road?

2. Obviously, make sure you have a spare tire for truck and trailer. Traveling with a caravan of buddies means no sharing. Everyone needs his or her own spares.

3. More than one good flashlight. Everything is tougher to do in the dark. Plus, the more light you have the safer it is.

4. Some sort of shovel to help clear dirt and grass away from the trailer or truck. Some road shoulders are soft and your rig is heavy. Expect a little sinking to occur. Don't get stuck.

5. Jumper cables, two sets if you drive a diesel. Trucks take a lot of juice to get going. Learn how to jump your truck.

6. Some sort of small tool kit. Not just to repair your vehicle. You never know what may occur. Pliers, hammers, screwdrivers and wrenches work a lot better than hoof picks for emergency tasks.

7. Gloves. A pair or two of heavy work gloves to protect your hands, also to keep dirt and grease out of your truck.

8. Make sure to have the proper size tools to remove tire lug nuts on both truck and trailer.

9. A blanket or piece of plastic tarp, just in case you need to crawl under the truck or trailer to check something out.

10. Flares or reflectors to line up behind your trailer to let drivers know you are working on your rig

11. 100 ft. of good rope for securing down anything that has been torn loose or broken. Sorry it does happen.

12. A very sharp knife. For cutting or carving up anything – rope, halters.

13. A first aid kit. For scraped knuckles, cut fingers, bashed-up elbows.

14. A Little soap and water and paper towels or baby wipes. To clean up after dirty work.

15. Overalls for roadwork.

16. Also, don't forget to have some hay and water for horses

17. just in case you are stranded for a long time in one location.

Even if you are physically unable to change a tire or you just don't want to, you should still have all these items on hand. If someone stops to help, they'll have the muscle and you'll have the proper resources.

Cellular phones are handy, although you shouldn't rely on them to work everywhere. NO matter what your cell phone company says, cell phones do not work in all places. CB radios can get attention from other drivers.

Finally, I'd like to talk about your personal safety. You can always repair or replace a tire and rim. If you decide to drive on and not stop, yes, there will be noise, sparks, and big sound. You may even damage your trailer. In the larger sense, your personal safety comes first – way out in front, not just a close second.

Then comes the safety and well being of your horses. Trucks and trailers are replaceable. You're not!!

Talk to your friends and listen to their stories about problems on the road. Compare this chapter to what these folks didn't do or didn't have - or things they could have used when the trouble hit. Safe traveling.

P.S. Toilet Paper!!! Need I say more… ☺?

Chapter Twenty-Six

"Electric Fencing"

When I'm talking to someone about temporary fencing or trying to keep their horses out of a particular area, when I say electric fence, their eyes seem to enlarge and they say "Oh no, no electric fence." I suppose they envision one of those late night movies, you know, the ones where some poor fool tries to climb a fence and just hangs there screaming and cooking like bacon in a pan. Well, livestock electric fencing is not that melodramatic.

Electric fence when installed correctly and maintained correctly can give you great peace of mind. Electric fence is very inexpensive and very easy to install. Also, it is an extremely easy operation to move this type of fencing about your property. Steel pipe, woven mesh wire or barb less wire all cost much, much more than electric fence.

Electric fencing can actually save you money and time, plus protect your horses. It is very likely that sooner or later you will come across a fence line that is old and tired, maybe even dangerous to your horses. Using electric fence in this situation can provide you with a multitude of conveniences, such as protecting your horse from the old fence line and adding much needed security to your pasture as well. Using electric fence can protect your horses from barbed wire, old trees, junk cars and deep ravines.

Now let me speak to the issue of a horse being shocked by electric fence. The power is very low, so low that a dog, cat or even a small child cannot be hurt. All it does is deliver anyone or anything a small gentle yet surprising little shock. I will admit when it takes me by surprise, it frustrates me a little when I get zapped. Yet I'm fine and so is every person and every animal I have known that has come in contact with electric fence.

Safety issues regarding electric fence are simple and inexpensive. When a storm is approaching, you should unplug your fence charger from any electrical outlet. Although now days most manufacturers install lightning devices in their fence chargers. I still say it's easier to unplug the charger than buy a new one if the fence is struck by lightning.

There are three types of electric fence charges: 110 V plug into the wall powered; 12 V battery powered; solar energy powered. Also, prices range from $35.00 to $300.00. Price is based on power and the distance of fence that can be supplied with a shock. Some have the ability to supply 50 to 100 miles of fence; others will barely cover a large back yard. Some can maintain a quality shock through wet weeds and tall grass; others will fail if anything is touching the wire. The charger that is best for you depends on the length of the fence and power source such as electrical outlet, battery or solar energy. Don't forget we are talking about horses not rabbits; that means you want a fence charger designed for larger animals.

Electric fence is similar to building with a child's building blocks. You can go any direction, any distance and change it or start over if you don't like what you just built. The key to electric fence versatility is that the fence does not have to hold weight. One or two wires can hold in horses. Think about that. If there is no power running through the fence a couple of little wires wouldn't hold in a pony.

The fact is that even the slightest contact that is made with the fence gives the electric fence its power and authority to hold animals in. You will never see your horses rubbing their manes or scratching their tails on an electric fence. I bet you have seen your horses rubbing and scratching on other types of fencing.

There are some down sides to using electric fence. First, once the horses learn about the fence, they usually stop grazing with in a 6" to 12" distance from the wire. That means mowing or cutting back grass and weeds that can grow tall enough to affect the shocking power of the fence. Along with this situation is small branches and tree limbs touching the wire. Obviously, by now you can see that if grass and weeds can affect the shocking power, your animals are going to be safe. Also the plastic or ceramic devices that hold up the wire onto a post are called insulators. If insulators become damaged or are not secured properly to the post, the wire will either fall down or be grounded out by the post. Grounding out is how your horses get shocked. The electricity is just pulsing through the wires. Something or someone has to complete the circuit to manufacture an electrical charge. Ever notice how birds can sit on a power line and never get shocked. That's because there is no grounding taking place. If there was a way for those birds to become grounded, those birds would become original recipe or extra crispy!!

Another down side to electric fence is the charger itself. Maybe it requires electricity. This means plugging it in. The cost is less than a 100-watt light bulb, although you need to protect it from horses, cats, birds and dust. Battery powered chargers need battery maintenance. Solar powered chargers are pretty much maintenance free. Solar powered chargers are also the higher priced.

I've told you about the chargers, safety issues, and also enough information to make a decent decision on whether you may want to install electric fence in your pastures. Now I want to write about the wire. The wire can be "galvanized wire", "stainless steel wire", or now we have "ribbon". Electric ribbon is a combination of cloth or plastic fibers supporting very thin wire mesh. A lot of folks like the traditional wire; others like the ribbon wire.

Metal wire is a whole lot tougher than ribbon. Metal wire also carries a better shocking power than ribbon, although ribbon is easier to repair. If the ribbon breaks, you just tie a knot just like fixing a broken shoelace. I really can't give you a reason to choose metal wire or ribbon. It's a personal choice based on what you want to see and also how much money you want to spend. Ribbon does tend to cost more than wire. Yet I have noticed in recent years folks are using more ribbon than ever before.

Installing or building electric fence is very easy. Electric fence does not require big solid posts. Metal T-posts or specially designed electric fence post will work just fine. Electric fence being so light and by design not having to physically hold in horses means posts can be spread farther apart than regular fencing. This, too, adds to the cost savings on electric fencing. I thought about whether to go into detail about actually building electric fencing in this book. I have chosen not to, due to the creativity and varieties of electric fence that are available in the market place.

Photographs and drawings would not illustrate electric fence fully. Actually, like most things in life, a hands on, right in your face look is the way to go. When you start to look at the way folks have set up their electric fence lines, you will see just how versatile it is. Also just how "willy-nilly" you can string it up. So after reading this chapter start eye-balling fence lines.

The last thing I want to address about electric fence is the ground rod. No matter what type of electric fence charger you chose to buy, when installing it, you will notice there are two places on the outside of the fence charger. Usually there are two screws or two wing nuts or two bolts. They will be labeled as follows:

1. Positive / Negative = Positive is where you hook your fence wire. Negative is where you hook your ground wire.

2. Fence / Ground = well after reading #1, not much for me to say here.

3. Line / Pole = Line refers to fence line. Pole refers to grounding pole or rod.

It is very important that the two never be switched around. Doing so will immediately burn out your charger. Also please remember to use a real ground rod. No T-posts or connecting the ground wire to an existing ground wires or any type of post. A real genuine ground is eight feet long and solid copper. These ground rods can be purchased at most hardware stores and lumberyards. Make sure when purchasing an electric fence charger that you also get the owner's manual or at least an instruction sheet on how to install it correctly. If it does not come with either of these, call the manufacturer for assistance.

Electric fence may not be for you. It does have many uses, although I will agree that electricity and horses does sound like a strange pair.

Hopefully this information will help to answer your questions about electric fence. Like you have so far and will continue to read in this book, ultimately, it's your choice. That's right folks. I'm saying it to you again. Your horse, your money, the responsibility is yours.

GASOLIN

Chapter Twenty-Seven

"Barn Burners"

Every once in a while, the weather turns cold and harsh. On days like that, most folks really don't want to mess around with their horses. That gives me the opportunity to write and work on radio show business. Seems like bad weather gives me the only chance to take a day away from horse work. Since I dearly love what I do, it's not really work at all. It's exciting and fun, not to mention very educational, to say the least.

The other day I was watching an old movie and the homestead in the movie was set ablaze by rustlers. Of course in Hollywood, there are no small fires, only giant catastrophes that burn everything in sight. Barns, sheds, the house, even the outhouse is gone. You know times are hard when the outhouse is gone too! Anyway, what does this have to do with your barn? Well, actually quite a bit; you would be surprised. When most people think about barn fires, it's like that old movie. There's a big, solid, dry wood barn. Piles of loose hay everywhere. Kerosene lanterns hanging on every post, not to mention the ammunition and gunpowder being hidden from the bad guys. Like I said, that's Hollywood.

So most horse owners think, through a process of elimination (no old dry barn, no piles of hay or straw, no lamps or lanterns hanging about, not much gunpowder around in barns these days either, or should I say, better not be any around?) that their barn is pretty safe from fires. And why wouldn't you feel safe? Metal roofs, metal walls, concrete floors, rubber mats, and treated lumber, open-air ventilation. Sounds good, sounds very fireproof. Like most things in life, things and situations are not what they seem. Let's cover fire safety for your barn.

I could spend days and days writing about different types of barns and sheds that have varied and potentially dangerous fire hazards. Instead I'm going to list fire hazard situations and fire hazard potentials. You can run through the list and see how your horse facilities compare to this list.

1. Exposed or faulty electrical wires and outlets, including extension cords and light fixtures.

2. Improperly installed or unsafe fans, heaters or water tank heaters.

3. Piles of old used feedbags or piles of old dry dusty hay, straw, and shavings.

4. The parking of cars, trucks, tractors in the barn.

5. Using a box stall or feed room to store lawn mowers, weed eaters, chainsaws, log splitters, any and all petroleum products.

6. Old outdated fuse boxes and circuit breakers.

7. Propane or natural gas heaters and pipes not being inspected on a regular basis.

8. No lighting rods or grounding rods installed or improperly installed rods on barn roof.

9. Dust and dirt collecting on and in light switches, electrical outlets, fuse and circuit breaker boxes.

10. Old worn out radios being plugged in and left on forever.

11. Electric or battery-powered fence chargers not installed properly or not cleaned and maintained correctly.

12. Refrigerators not being cleaned and electrical cords not inspected regularly.

13. Window air conditioners not being inspected or serviced regularly.

14. No fire extinguishers or smoke alarms.

15. Alleyways, doors and windows of the barn blocked or obscuring ventilation and natural air flow.

16. Insulation and padding not properly installed or making contact with electrical wiring or outlets or fixtures.

17. Overloading of circuits, breakers and outlets with the use of too many electrical devices.

18. Adding new electrical wiring or fixtures and outlets to old outdated wiring and junction boxes.

19. Connecting electrical power to other smaller sheds and buildings from the barn without properly increasing wires and breaker boxes.

20. Old or faulty yard lights and night-lights.

21. Improper installation or no inspection of foal monitors, cameras and sound equipment.

22. Horses being able to play with or chew on electrical wiring and outlet fixtures, plus any fans, heaters or radios.
23. Mice and rats chewing on wiring, building nests in outlets, fixtures and fuse boxes, just being very destructive.

24. Birds, cats, dogs, skunks, possums, raccoons, squirrels, chickens, ducks, geese, or goats basically doing what the mice and rats can in #23.

25. The most important thing to remember: NO SMOKING IN THE BARN!

After going over this list, you should have a good idea of how safe or how unsafe your own horse barn is and the potential of fire danger. Hopefully by reading this you can prevent what is one of the most traumatic situations for any horse owner. After reading this chapter, I hope you will check out your barn for something I didn't mention that can cause a fire: saddles, tack, blankets, pads, even liniments, hoof treatments, medicines; just about anything can burn or ignite if the conditions are right. Remember even the old outhouse can go at any minute!

WESTWAY
CONVERTER
HORSE PELLETS
4X4
OFF ROAD
TRAILBLAZER

Chapter Twenty-Eight

"Feed Stores"

You would probably never expect your local feed store to be a great place to learn about horses. Most of the time we drive up, jump out, and rush in and buy what tack, hay, or grain we need.

Surprisingly there can be a lot more at your local feed store than you may be aware of. Almost all of the feed companies – most of the companies that make wormers and drugs – also provide lots and lots of good horse information.

There are videotapes, books, pamphlets, even cassette tapes and CDs on just about every subject for your horse. The great part is, it's usually all free. That's right. No charge. Just good information on yearly shots, worming, colic, losing weight, even how to put weight on your horse. I've even seen a tape on how to use rub downs and liniments. Sadly, nobody seems to know these freebies exist.

I was in a feed store yesterday and I noticed a box with about 100 VCR tapes in it. The tapes were from a major feed company; the subject was caring and feeding for a new foal.

What a great tape to have and it was free! I had a conversation with the storeowner as to why the tapes were just lying around. He stated that he felt funny about talking to people about the tapes. He did not want to bother the customers. So I grabbed a tape, stood in the middle of his store, raised my voice and mentioned the tape and how the tape was free. I passed out eight tapes right out of the chutes. The storeowner called me today and said that he was just about out of the tapes, and that his customers are also asking for anything else he may have on hand. Well, sometimes things just go perfect for all.

The point is, folks, you probably are going to have to speak up about this free information on your own. Also, feed stores do get offers from companies and sales reps all the time about informational stuff, but I've found that the store often declines these offers. The storeowner figures nobody will want the stuff.

So don't be shy. Speak up and ask for any and all the information your local feed store can get you. I bet you'll be pleasantly surprised as to what the store has to offer, and also what you can learn.

See you at the feed store.

P.S. I'm the one in the back corner always reading feedbag labels and horse care product information. ☺

Notes Page

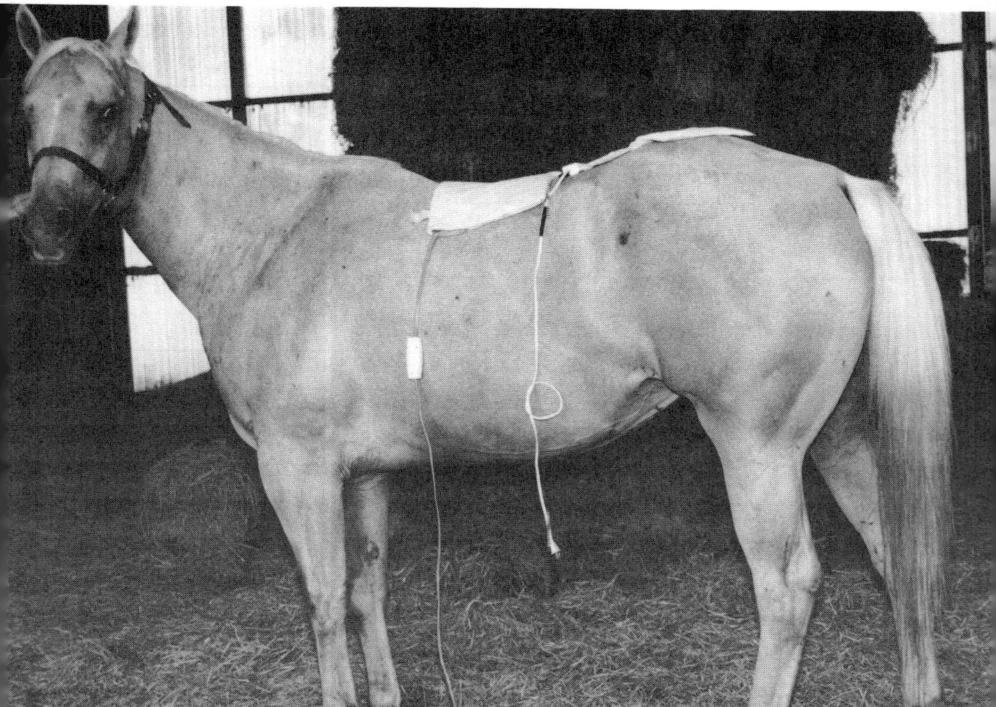

Chapter Twenty-Nine

"Horse Blankets"

It seems to me, no matter where I go or whom I talk to, if they own a horse, blankets come up in a conversation. Not just blankets in general, but what type, what size, brand names, also turnout sheets, fly sheets, coolers, and hoods. Most folks get confused and end up spending way too much money.

First, let's discuss the horse. The perfect air temperature for a horse is 50°, so if it's 40° to 50° any type of covering is probably not needed. Also given a fair chance, most horses will grow a nice thick winter coat to stay warm. Show people don't particularly care for a shaggy horse. So keeping a horse blanketed will slow down its hair growth.

Now, about barns and shelters. If you have a barn or pole shed, when the weather is bad your horse always goes in it. The horse stays out of the wind, stays dry. A cover is probably not always necessary. Actually in most instances people overuse and misuse blankets. I realize that sounds a little funny; yet it is very true, and happens often. Blankets, even sheets, are often put on horses before they actually require them. For example, it's between 30° and 50° and a horse has a nice warm blanket on. The sun is out, not much wind. The horse is hot, possibly even sweating under the blanket.

The next thing in this domino effect is dirt. Not just dirty, soiled blankets, but also dirty horses. A horse should have a bath and a good grooming before you put a blanket on, especially if you intend to leave the blanket on for days. With the horse being a little hot, a little sweaty, and dirty, you get the makings of skin bacteria fungus.

Blankets have a very good purpose in a horse's life as long as they are used properly. Remember folks, any type of covering also has some dangers attached to it. Horses love to roll, scratch, and groom each other. If a blanket does not fit properly or is not secured just right, disaster can strike at any moment. Blankets too big can twist and turn on a horse's body. Also oversized covers can hang down too low, which can interfere with the natural ability of the horse's skin and hair to breathe and move, not to mention it's just plain uncomfortable for the horse. Keeping rips, tears, and most importantly, all straps, buckles and snaps in good working order is key to safety and ease of use.

Many times people buy more blankets and covers than they really need. Just because your barn buddy has everything ever made to protect a horse, does not mean it's right or that it's even appropriate. Usually one good, all-around blanket is fine for bad weather. Maybe a light fly sheet or nylon cover for a cool breeze. Your area's climate and the type of facility your horse uses to stay out of the weather really should be the two biggest deciding factors when choosing coverings.

As far as the different types of blankets, rain covers, coolers, fly sheets, turn-out coats and hoods, there are just two many styles and types to list. You can find the previously mentioned tack in most tack catalogs and tack shops.

Don't forget to check out your barn buddies' blankets to see how each one works and also to see which ones are needed and which ones are not. Remember this, the more you alter and make adjustments to a horse's day-to-day life, the greater the odds are of something going wrong.

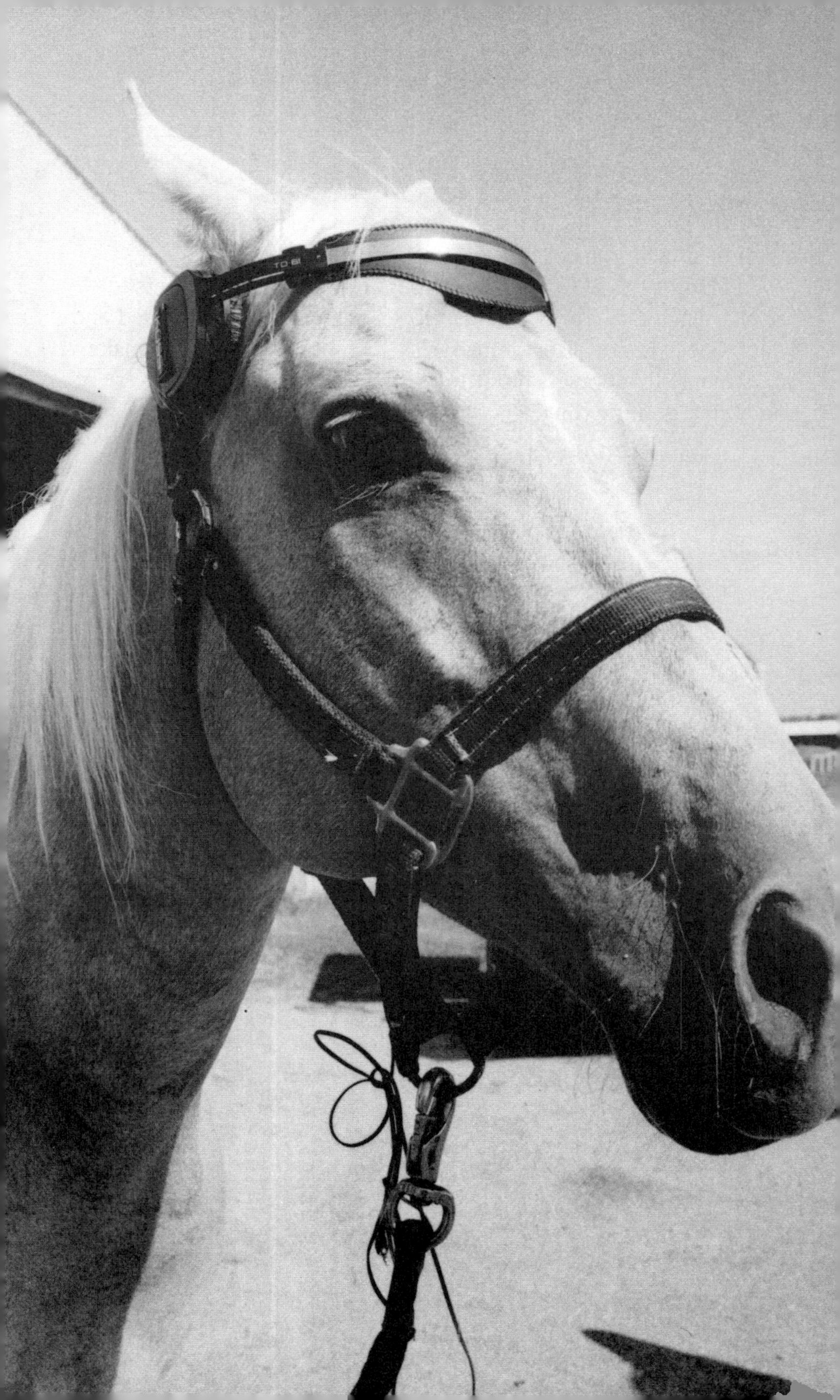

Chapter Thirty

"Horses Can Hear"

I'm just not amazed much anymore by the beliefs, wives tales, or just plain lies that horse people think are true. Over the course of time, I'll cover all of these pony poop facts.

Right now, I'd like to discuss your horse's hearing with you and I'll address some of these pony poop facts. Any horse, any breed, any age, or any sex has just about the same hearing acuteness. Sure, there are some basic variables. Take genetics, maybe Mom or Dad was hearing impaired due to a condition at birth, or because just one of millions of genes passed through the ages from generation to generation. Remember, not all genes are bad. Registered a horse lately? Genetics plays a big role in color, size, and speed, even personality, hearing, sight, not to mention soundness of body.

Most research done on the ability of a horse to pick up and associate sounds is limited. Yet some facts do come to light in every research project. A horse seems to have the ability to pick up sound as well as a very well trained hunting dog. I've tested my horses against my own dog. My dog comes from a long line of champion field trial Labrador Retrievers. Yet, since she is my best friend and a day-to-day companion, about the only thing she hunts are snacks and shady spots to take a nap, while I work on horses.

For the hearing test, I took a small bell and connected it to a car battery. Then I hooked up a remote control to signal the bell when to ring. I put this device way out in my pasture. (I know, too much time on my hands.) First I started with my dog. I started so far away I could not hear the bell ring. At first my dog just casually looked, then dismissed the sound. I marked each spot with a dot of spray paint in the grass. As I moved closer, I could tell by her ears, eyes and body language that the bell had her attention. For a long distance I couldn't hear the bell yet.

Next I took each one of my horses individually along the same course. Again marking each distance. Every horse reacted slightly differently; yet close enough to show a pattern. Then I got my neighbors involved in this. Results were similar except for three horses. One was over 25 years old. The other two had been used on a lot of hunting trips. The older horse was really no surprise to any of us. The two who had been on hunts really had us talking. Apparently, long and regular exposure to gunfire had taken its toll. We realized that their hearing was diminished enough to affect their behavior.

Let's think about this for just a second. A horse's most important sense, besides smell, is hearing. A horse is designed to hear trouble first and then see trouble. Finally, the horse hits high gear and runs from trouble. After the remote control bell, I tried whistles, horns, radios, walkie-talkies, televisions, and even a tape recording of a wolf pack howling. When I played the wolf pack tape the responses were even more dramatic. After I caught all the horses, fixed the gate, fed the dog, I figured maybe the wolf pack tape was not such a good idea.

Well, I do try to learn as much as I can but, like anyone, not always in the easiest or smartest of ways. My dog's hearing was better than any of the horses. Not by much, but she did have an edge. What does this mean for you, the horse owner? Just think about the following list and you'll know where I'm going with this chapter.

- Truck exhaust systems
- Radios in barns
- Loud speakers at arenas
- Barking dogs
- Gunfire
- Fireworks
- Yelling at your horses
- Slamming trailer doors
- Thunderstorms
- Construction sounds
- Air conditioners
- Generators
- Electric hair trimmers
- Horseshoeing sounds
- Tractors
- Airplanes

Well, you now get the idea about how sensitive a horse's hearing really is. Limiting constant exposure to sound is very important to the overall comfort of your horse. Not just the comfort, but maintaining good hearing ability as well. Try snapping your fingers, or having someone make even the softest of sounds out of your horse's eyesight. Notice the ears rotate towards the sound, the body positioning itself for panic and escape, or the horse moving closer to investigate the sound.

Keep sounds to a minimum or, if at all possible, limit the time exposed to the sound. Are radios turned on 24 hours a day, seven days a week harmful or disturbing to a horse? Think about what you've read. Also try some tests of your own and you decide.

Finally I'd like to add that I've thrown the wolf pack tape away. After the goat incident, I really needed to make some adjustments.

Chapter Thirty-One

"Cribbing"

Well it's not just raining; I feel like the flat rock that a cow is unloading on. I'm sitting here on a bucket waiting for my clients to come home so I can float their horse's teeth. The clients call. They're running late. I've played with the barn cat and snooped around the tack room. Hey come on! Everybody likes to snoop around other people's barns. You never know what you'll find. I've always thought snooping around someone's barn and tack room is like checking out the medicine cabinet in someone's house. Nobody admits doing it. Nobody thinks it is a nice thing to do. Yet just about everyone does it one time or another. I'm looking for paper.

I really never thought about this. I suppose you just figured I sat in front of a computer or typewriter writing this book. That is not how this book was written. Some of it was written in my house sitting in my chair with a clipboard, using binder paper and an ink pen. Most of it was on feed sacks, note pads, candy wrappers or paper towels while someone was catching his horse or I was waiting for a client to come home. A lot was written over the years; all I had to do was to put it all together.

I've been waiting for about 30 minutes. It's pouring out. Normally I have time between appointments; usually, though, I have one appointment right after another. Today, two or three all day, a mini vacation. While sitting here, I started to notice a scraping sound along with a gasping of air. The telltale signs of a "cribber".

Cribbing is when a horse takes his upper incisors, those front upper teeth, hooks them on a post, board, or gate and stretches his neck out and chews and sucks in short puffs of air. Cribbing is a nasty little habit. The keyword is "habit". Horses are extremely susceptible to habits, good and bad. The rapid intake of air causes a horse to experience a mild form of hyperventilation. This gives the horse a kind of mild "high". Along with chewing action, it doesn't take a horse long to become an all-day every day cribber.

Electric fencing, wood painted with spices, and anti-cribbing paints have mild effectiveness. Cribbing collars prevent a horse from extending their neck out to crib. If used properly, a horse can still eat and drink while wearing the collar. I have seen horses continue to crib even with the collar on!

Horses generally chew wood for a couple of reasons. First, boredom; horses were built to walk 10-20 miles a day, grazing casually for 12 to 16 hours of that period. Considering the number of horses that are placed in stalls or small pens, I think it's amazing more horses don't act out due to boredom.

Another reason for cribbing is a nutritional problem. Everybody thinks about protein, yet vitamins and minerals are more important than protein. If for some reason your horse's diet is lacking in nutrients, the horse will search out a way to find them. Hence, chewing on wood, trees, gates, posts, farm equipment, even horse manure.

Some researchers feel that cribbing can be hereditary, and passed down to foals. Personally, I'm not sure about that. What I am sure of is that a horse that cribs is a horse that can literally eat your barn away. Even worse, small wood and debris in extreme cases can build up in a horse's stomach. This can cause internal bleeding into the stomach. Sometimes blood can be seen in the horse's manure. Or an internal infection can start due to the wood particles affecting the stomach lining. The products on the market that are designed to remove sand from a horse's stomach and intestines has no effect on wood particles. This is what makes cribbing such a risk for your horse's health, not to mention your barn or fence.

If you notice any of your horses cribbing, do everything you can to stop it A.S.A.P. Make sure to consult your veterinarian before applying a cribbing strap to make sure that it's a proper fit. There are several tack accessory companies manufacturing cribbing collars. Each claims to be the very best. I personally have seen just about all of them work. More importantly I have seen them all fail as well. The ability of a horse to turn cribbing into a life long habit doesn't take very long. Please believe me when I say it is a hard habit for a horse to break. Unfortunately a lot of horses never do break this habit. Horse owners can spend years and years going back and forth battling a horse with a cribbing problem. It is all these reasons why you need to keep a sharp eye on your horse facilities and your horses. If you bring a new horse onto your property, verify that it is not a cribber. A horse that cribs habitually can train other horses to crib just by example. Young horses under the age of two years can change the growth pattern of their teeth by daily cribbing. Young horses' gums and teeth are growing. Cribbing can cause the gums and incisors to grow and mature in a deformed state. This can lead to grazing and eating problems. Also, this abnormal tooth alignment can alter the horse's muzzle and upper lip. Injury and permanent damage to the lips, gums and tongue have been associated with cribbing. Horses are designed to graze and eat grass, not to chew down trees and build dams!!

As for the snooping around in the barn, I wrote this on a roll of paper towels I found in the grain room. I needed some kind of paper to write on. My clients just laughed and stated that only I would sit on a bucket in a horse barn and write a book on a roll of paper towels. Well, as far as the paper towels are concerned, it was easier than toilet paper. Please, no jokes. I probably have already thought of them. As for where this chapter was written, what better place to write about a horse cribbing than four feet from a box stall with a horse doing it?

I'm going to get busy here. By the way, my clients wanted the paper towels back for a souvenir. I considered it a very great compliment when they wanted me to autograph the roll. I have heard through the grapevine that the roll of paper towels is in a plastic case in their home on display. Hopefully not the bathroom; mistakes happen!!! ☺

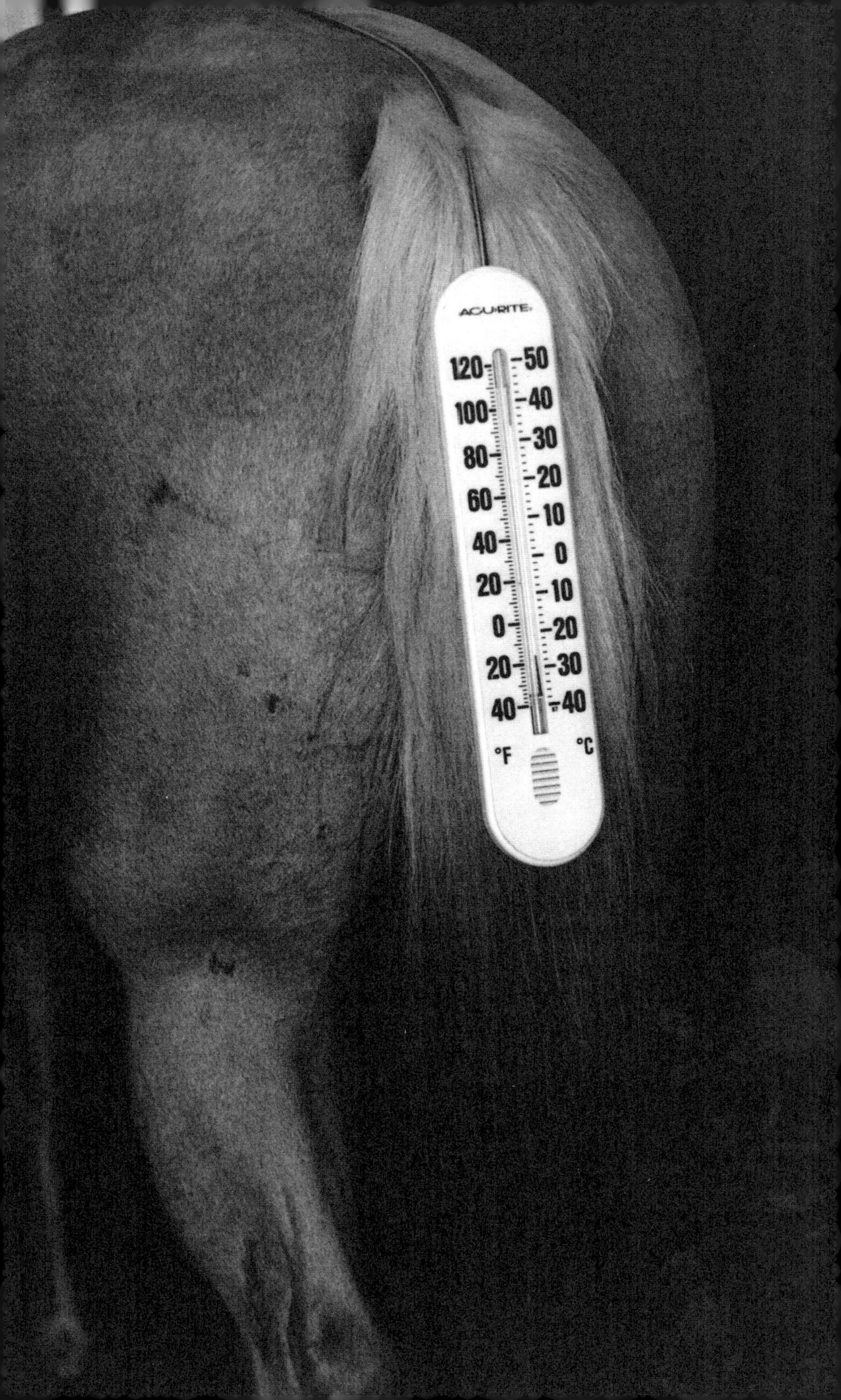
ACU-RITE
120
100
80
60
40
20
0
20
40
50
40
30
20
10
0
10
20
30
40
°F
°C

Chapter Thirty-Two

"How Hot is Your Horse?"

Taking your own temperature when you're sick is pretty easy. Yet when we have to take our horse's temperature, a lot of us get a little frustrated and intimidated by the task. The actual process of taking a horse's temperature is not so hard to understand. You take an animal rectal thermometer, slide it in the rectum and wait two to three minutes. Pull it out and read the number. Sounds easy, right? Well, it can be, or it can be a disaster.

A lot of horses will kick and move around if you go messing around with their backside. Also, believe it or not, a thermometer can actually be retracted, so far back into the horse by a muscle reaction that you can't get it out. I've seen this happen. What a mess.

Okay, now that you have an idea of what can go wrong, let's talk about what to do to at least try to make sure everything goes right.

First the thermometer: buy one that has a small eyelet or hole in the end of it. This hole is to thread a string through to prevent the horse from drawing the thermometer into the rectum too far. Or you can get one of those fancy digital units, but unless you are a vet or you plan on having a bunch of sick horses, the digital one may not be worth the money to you.

Now let's go step through the proper and safe way to check a horse's temperature.

1. Have someone control the horse at the front end with a halter and lead rope.

2. Rinse thermometer with soap and water or alcohol to disinfect it.

3. Read thermometer and shake it down to at least 65° to 70°.

4. If applicable, attach string or line to thermometer to prevent losing it.

5. Apply lubricant on thermometer to help to slide it into the rectum.

6. Check the temp again to get an accurate reading.

7. Lift the horse's tail; slide the thermometer in gently and smoothly. Also stand to the side to prevent getting kicked. Have the string or line securely in your fingers.

8. Once the thermometer has been inserted, leave it in for two to three minutes. The longer, the better.

9. Pull the thermometer out quickly, yet gently. Watch out for kicking.

10. Read temperature as quickly as you can.

11. If a fever is suspected, wait 10 minutes and verify by doing another temperature check.

12. Obviously clean and store thermometer.

These temperatures are averages and should not be relied on as perfect:

Adult horse: 100° = normal
Young foal: 102° = normal

The best procedure is to take the animal's temperature, observe its overall health and condition, and then contact your veterinarian with your findings. Good luck. Hope this chapter will come in handy.

P.S. I do know of a person who actually had two fingers bit off trying to take a horse's temperature. That's right – the right intentions, the wrong end. ☺

Chapter Thirty-Three

"Spreading Viruses"

One of the most troublesome situations for any horse barn is the spread of colds, flu and strangles. Even certain skin funguses can be contagious. You obviously cannot completely eliminate diseases in your barn. Yet you can go a long way in prevention, and cutting down on one sick horse, causing the whole barn to come down with it.

The most common way of spreading viruses is as follows:

1. A sick horse's virus is passed to other horses breathing the same air. This happens often in a closed barn.

2. The discharge from a horse's runny nose is left on feeders, buckets, walls, doors, bedding and wherever they drink water.

3. Horse tack such as halters and lead ropes used on more than one horse can spread viruses.

4. Shared grooming brushes and hoof picks can spread funguses, thrush and several types of skin bacteria.

5. Brooms, shovels, manure forks, wheelbarrows – any tool that comes in contact with a sick horse's bedding - can potentially spread the problem from one stall to another.

6. Shared syringes or needles, improper cleaning and sterilizing of floats and farrier's tools.

7. Not taking proper steps to quarantine a sick horse.

Now that you have a good picture of how easy it is to spread viruses, let's review how to protect the rest of your barn from one sick horse. Try to have one stall that is separated from the rest of the horses. Leave a stall open on both sides of the sick horse if possible. Put up plywood or plastic sheeting to slow down airflow of viruses. Install a fan to blow air out of the stall; have the fan blow out a window. Also, when disposing of the dirty bedding, make sure to keep it away from other horses, possibly removing the dirty bedding from the property. This is important when you have a lot of sick, young horses.

When it comes to strangles, especially when there is draining taking place, halters can pass the virus from horse to horse. Try to have a certain set of grooming equipment, feeders, and tack that is used only on the sick horse during the outbreak. Plus, probably the most fundamental mistake made is not following the veterinarian's advice on quarantine and medication. Do what your horse doctor tells you. Let the drugs do their job!

Another way to spread disease is thinking a horse is cured sooner than it really is. In most cases, after the nasal discharge and coughing stops, you need to continue the medication another four to seven days. Also, after all the drugs are used up, you still need to keep that horse away from its barn buddies for five to eight more days to prevent relapse and any contamination taking place.

Example:

Day #1: At morning chores, horse coughs and/or has nasal discharge and fever. Horse is off feed.

Day #2: Fever gone from medication and other drugs combating the virus.

Day #3 through #8: Medication treatment continues.

Day #9: Medication stops. All symptoms gone.

Day #10 through #18: Horse is back on full feed, acts normal and healthy yet can be susceptible to relapse or pass virus on to another horse.

Day #19: Horse goes back to herd or barn.

The schedule and length of treatment and quarantine will depend upon what's wrong with the horse and how severe the sickness. If possible, have a small building or pen specifically set up and segregated from the rest of your horse facilities, just for contagious horses.

This area can also be used as a foaling area, a spot to keep a horse when waiting on a Coggins test, maybe a temporary stall for someone just passing through town. Over the years, I've always found a use for such a place. You can even wean a foal in it. In the end, the faster you spot a sick horse, the quicker you get it started on medication. The farther you keep it away from other horses, the better off everyone will be.

Always discuss all the options and best way to handle the situation with your horse doctor. As we all know, every year something makes its way through the barn. Hopefully the virus won't be too strong or last very long when it does occur. Unfortunately as I write this, West Nile Virus has crept its way into Texas and Oklahoma!!

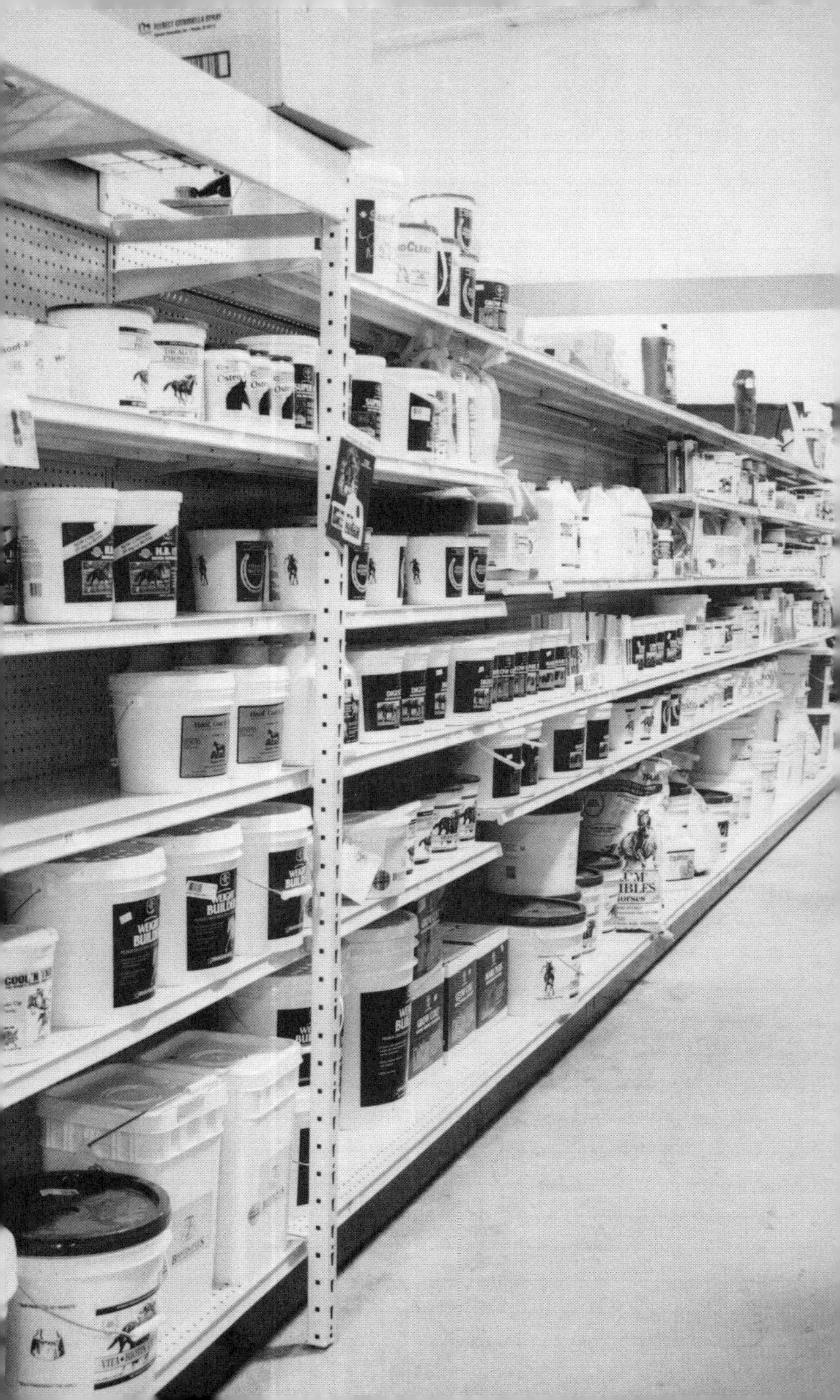

Chapter Thirty-Four

"Fly Spray Allergies"

Fly sprays. Okay, okay, I completely agree. Most of them are just junk. Someday someone will come up with a spray that actually works for more than an hour and doesn't cost $100.00 a bottle. Now that I've slammed the fly spray industry, let's get to work dealing with what can happen. It's actually quite common for fly sprays to affect your horse's skin in a negative way.

Allergic reactions to sprays come in many forms. It can be a loss of hair or just a loss of hair shine. It can be small bumps, almost like bug bites. The horse's skin can become sensitive to touch, just irritated from the chemicals. Runny noses, tearing eyes, change in disposition can also be related to fly spray reactions. The biggest misconception is that homemade fly concoctions are safer than store bought types. That is a falsehood. Chemicals are chemicals. Also a lot of homemade fly repellent recipes mix items together that were not designed to be put together, especially on a horse's skin. A horse's skin is extremely sensitive and was really not designed for any fly repellent at all.

As in a lot of chapters you're going to learn that we (by we, I mean horse owners) put horses in places and circumstances that are easy and convenient for us yet put the horses' health in direct jeopardy. In a horse's natural habitat, flies are few and far between. Stables, paddocks, garbage cans, manure piles: these are not natural habitats. Yet I just described the areas most commonly found around today's horses. So given that our horses are around flies, fly sprays are needed. This is the problem. In an economy like ours, competition is key. Sprays are becoming stronger and stronger. This can lead to more sensitive skin. As the custodian of your horse's well being, it's up to you to be aware of what a spray does to your horse's skin.

When trying a new spray, test it first on a small area. Look for sensitivity, bumps, and hair problems. Also see if it really keeps flies away. If you find a reaction, first rinse the horse off with plain water from the hose. Obviously don't apply any more fly sprays. Even switching brands after one brand has failed can cause you trouble.

Contact your local horse doctor for advice, and in severe cases seek proper medical attention. Keeping horses out of direct sunlight can help relieve fly spray irritation. There are other ways of keeping flies off horses: face masks, full body fly sheets or fly tags attached to halters.

Remember, fundamentally, no matter what it is or what it is for, anything from shots to wormers to fly sprays can cause allergic reactions in a horse. Don't stop using them. Just keep in mind to be cautious and read all manufacturers' labeling and instructions. Keep a sharp eye out for problems.

Chapter Thirty-Five

"Reading Horse Manure"

I'm sure just about everyone has talked about what they would do with three wishes. Every once in a while, it seems I'll be having that old conversation with someone. Three wishes – fame, fortune, or magical powers. Myself, I always wish for the same three things. The first wish has always been the ability to talk with horses. It would make my life so much more exciting and I could do so much more for a horse's well being.

Unfortunately, my three wishes haven't come true yet. That means when dealing with a horse, I have to use every sign and signal possible to understand and determine a horse's health. One real easy and basic window that you can look into to ascertain a horse's health is the horse's bowel movements. That's right, the pasture peaches, stall apples, horse manure. Those little green aromatic balls of digested grain and hay.

Although most horse owners feel that even the mention of horse manure is a prelude to doing chores, you can do a lot more with it than just forking it, shoveling it, or hauling it away. I'm not talking about granny's garden either.

Checking your horse's droppings on a regular basis can offer lots of information on your horse's current health status. This involves inspecting it when it's fresh, which means, if possible, you want to inspect it as soon as it hits the ground. There are several things that can be brought to light just by poking and breaking up the manure balls. Here is a list of situations and signs that can be noticed by inspection.

1. You want your horse's manure to be in shape of balls that partially break apart when they hit the ground.

2. The balls should be soft and very moist to the touch.

3. Balls that are hard and dry mean the horse may not be drinking enough water.

4. You may spot certain types of worms in manure, especially after worming the horse.

5. Undigested food such as corn, oats and pellets may mean your horse's teeth need floating.

6. Loose or runny stools are an indication of rich diet or worm or viral problems.

7. An average horse weighing around 1,000 pounds should have six to ten bowel movements in a 24 hour period, which may weigh approximately 10 to 20 pounds total.

8. Any sign of discoloration such as blood or mucus can be a sign of a serious health problem.

9. Pasture grass, or the type of hay being consumed can dictate the overall color of the manure.

Okay, I admit this particular chapter is not very appealing, yet you must realize how important it is that you have a very good understanding of your horse's metabolism. Unfortunately, none of us can communicate with our horses. The great thing is we can learn to observe all the body signals and characteristics we can, to achieve as much insight into the horse's life as possible. One of these insights just happens to be knowing and understanding your horse's manure.

ROLL-ON
WIPE
BRONCO
BRONCO

Chapter Thirty-Six

"Feeding Supplements"

When I was a small boy, my grandma used to make me take a vitamin every morning with breakfast. Also she made sure I had a balanced meal all the time. Extra helpings of vegetables and not too many sweets as well. My grandmother took wonderful care of me.

There is a fine line with a horse. Supplements have their place. There are products that build up the blood. Some concentrate on the joints and cartilage, others help to promote hair and hoof growth. If you list each part of a horse, somebody makes a supplement for it. The real trick is not just purchasing the correct supplement, but knowing when your horse really requires some extra intake of vitamins, minerals, proteins and other types of chemicals.

Please don't misunderstand me. Horses are very lucky these days. Supplement research and development has really come a long way in the past few years. There are some fantastic products on the market to help your horse.

Generally speaking, you really need to understand your horse's nutritional requirements before you start dumping little plastic scoops of powder on your horse's feed. Most horses get a good supply of protein, yet vitamins and minerals take a back seat.

A lot of times a horse owner will state how well their horse is doing since starting them on a supplement. They can definitely improve a horse's health, as long as you use them correctly. No matter if your horse needs its feet to grow a little faster, needs a longer mane and tail, more muscle mass and energy, you need to consider these important facts:

Fact #1: Horses are foragers; they are by nature designed to eat grass not processed grain.

Fact #2: Most horses are not fed what they need. They are fed what the horse owner likes or can afford to buy.

Fact #3: People consider hay as a filler and grain or pellets the main course or diet.

Fact #4: Most horses are fed the same amount and type of feed their whole life. As a horse matures and changes, their food should change as well.

Fact #5: People think that they should see instantaneous results after just one day of feeding a new supplement. It may take a year for results.

Fact #6: In a lot of cases, as we have four seasons of weather so should a horse have a variety of feeds to react to and coincide with the year's seasonal changes.

Fact #7: The majority of horse owners do as their barn buddies do. It's really the blind leading the blind. Remember each horse has its own diet requirements.

Fact #8: Every product manufacturer states their product is the best. This is a perfect reason why you should read product labels.

I can barely scratch the surface of this subject matter. As to what and when to feed a supplement, I cannot say in this chapter. What I can say with great sincerity is this: consult your horse doctor, ask questions at your local feed store, and really investigate your horse's needs. What I mean is, find out what is lacking in your horse's diet. You may just need more of what you already feed or a better quality feed. If your horse has a specific condition that may improve by feeding a specific supplement, then choose wisely and really feed as close to a perfect supplement as you can.

I realize this particular chapter may leave you with more questions than with answers. My response to that is, "great!" Great questions lead to great answers.

HORSE RATION

Chapter Thirty-Seven

"Slow and Steady"

After a few years of radio show questions, I've decided that a chapter on the basic way to change a horse's diet is needed. Switching feeds is a part of just about every horse's life from young foals to weanlings, yearlings to young adults, adults to seniors. Whether it's for performance or just a nice well-balanced maintenance, feed could be the way to go.

In a perfect world, a horse owner would be able to go to a feed store, buy one type of feed that would fulfill a horse's nutritional needs for the entire horse's life. No matter what the horse's daily activity is or whether it's a mare, stud, gelding, pony or draft horse. We're not in a perfect world. We are in the world of horses. Remember a phrase called "horse time". That world is one of change, compromise and, most importantly, flexibility. Now that you have a sense of what we all face, let's get technical, and discuss how to change a horse's diet with as little chance of risk and problems as possible.

1. No grain to a grain diet. When a horse is not used to grain, or has not had grain in a while, you need to start the grain very slowly. Start with no more than 1/4 of a pound or less for at least three to four days. Then gradually increase every two to three weeks to complete.

2. Switching from one feed to another. Switching from one feed to another should also be done over a period of 12 to 16 days. Start by blending in just a small amount of the new feed into the old feed. Do not change the amount of the meal, just the ingredients. As the days go by, you should gradually and consistently be gaining in new feed and lowering the old feed until you're feeding 100% new feed.

3. No pasture to pasture. As with grain and bagged feeds, a horse also needs time to adjust to free grazing on a pasture. This is best accomplished by giving the horse limited access to the grass. Start by letting the horse graze for two to three hours a day for the first week. Then every couple of days, increase grazing time. Try to increase grazing over a three-week period before turning the horse out full time.

4. No baled hay to hay. By now, after reading all this, you should get the idea as with everything else, start gradually by feeding small amounts of baled hay. Try to feed the hay in small amounts twice a day.

Increase the meals slowly over a period of 12 to 14 days, finally reaching full hay diet.

5. From one hay to another. You want to mix the hay slowly over a 7 to 10 day period. While you mix the hay, try to make sure the horse does not sort out one type of hay. This may be tough, especially if you're going from grass to alfalfa. Whichever hay is sweeter and tastier will be the most popular so make sure to monitor the process closely.

With any change, you need to keep in mind that it needs to be done slowly and with patience. A large percentage of colic and founder cases occur during a feed change. Horses have very delicate digestive systems. Horses are not as tough as cows or hogs when it comes to feeds. Also the age, sex, breed and size of a horse will dictate the amount of feed you use. Make sure to read all labels and manufacturer literature when changing a feed. Don't forget to consult your local horse doctor as well. Go slow, ask questions, and things should go just fine.

Chapter Thirty-Eight

"Spring Pasture"

One of the hardest things that seems to happen every year, no matter how much you plan or how well organized you may be, there is a race between running out of hay and having green pastures. Obviously where you live does affect spring thaw and when pastures will be ready for grazing.

Every spring, thousands of horses all over the country become foundered or colicky or both due to the exposure to spring grass too soon, and too much as well. The age of the horse really isn't a factor in this situation. All horses, no matter age or breed, are susceptible to grass foundering or colic.

New spring grass is lush, wet and very tasty to any grazing horse. Why not? Being fed the best of hay for months pales in comparison to new spring pastures. The trouble seems to creep up on a horse so slowly and subtly that you've got problems before you realize it.

The most fundamental rule, if not a law, when dealing with horses is any and all changes to a horse's daily diet need to be done slowly and gradually. I've written in other chapters about how switching feeds, or changing from one type of hay to another needs to be done slowly. Putting a horse out on pasture needs the same care and the same rule as any other feeding adjustment.

Here is a general plan for putting horses out on pasture.

1. Start with two or three hours of grazing time. As you increase grazing time, cut back and reduce hay and grain rations.

2. The process of grazing time should increase every three to four days. Make sure the horses are drinking sufficient amounts of water.

3. Check the horses' stools to monitor bowel movements. Stools should maintain firmness and moistness; loose stools or diarrhea means pasture is too rich for horses, so slow grazing time down.

4. The slow and steady rise in hours of grazing time should take 2 ½ to 3 ½ weeks to complete. Remember the more pasture grass a horse eats the less demand there is for hay.

Where you live and the quality and type of pasture make a difference as to how fast or slow to ease a horse onto pasture grass. For more detailed help specifically in your area, contact your local extension office or county agent; a short visit with your horse doctor doesn't hurt either.

Hopefully, this chapter and others I've written dealing with changing and adjusting a horse's diet will help you in preventing or at least diminishing the risk of founder and colic to your horses. Unfortunately, even under the best of conditions, things can go wrong. As a horse owner, I guess I really did not have to say that; owning a horse is proof of that on its own.

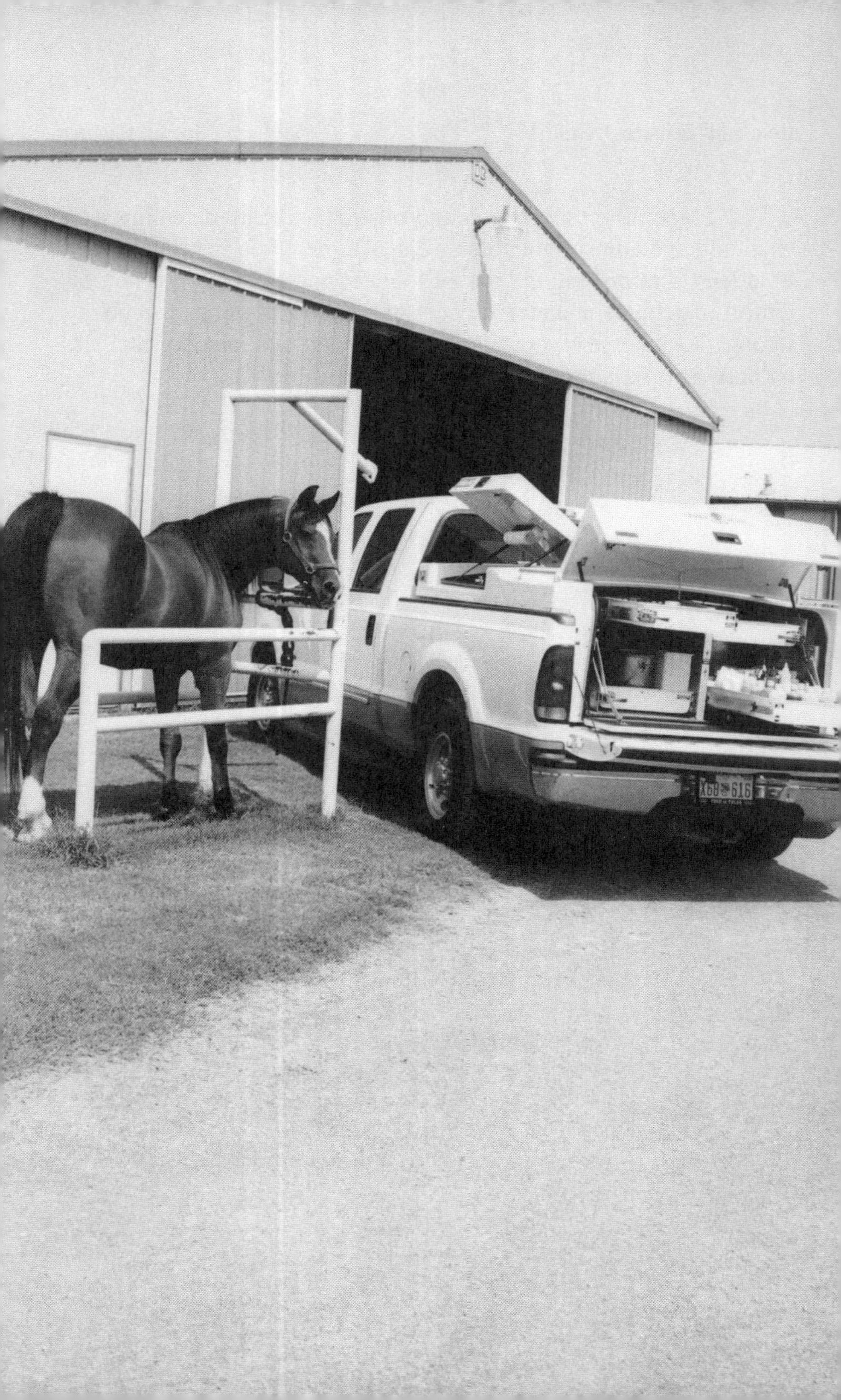

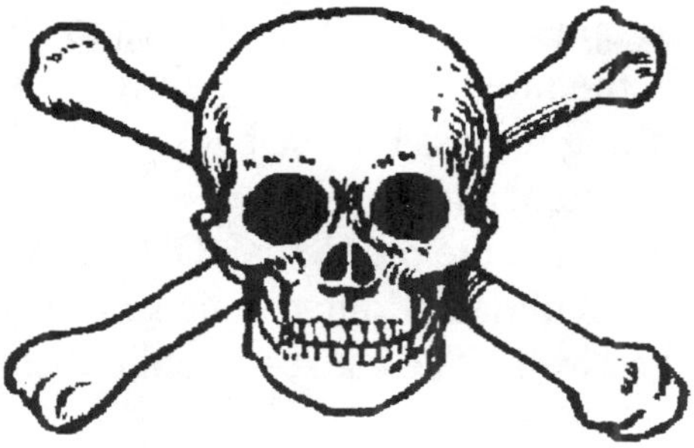

Chapter Thirty-Nine

"Colic"

Whatever you do – whether you're patrolling the streets on a saddle, checking fences in the pasture, looking at your stopwatch for the fastest index, turnin' and burnin', dallying your rope, riding the rail, jumping it, or just enjoying the day on a trail – the number two killer of horses is lurking at every turn.

More horses die from colic than any other injury or disease. Only lightning kills more horses. Colic comes in many forms. Is it hereditary? Depending on whom you talk to, you'll get varied responses to that. What I'm going to do is talk about different kinds of colic, signs and symptoms, and what to do and what not to do.

First, what is colic? Colic is a breakdown in the gastro-intestinal tract. Here's what occurs with a mouth full of food: The horse chews it. The teeth are supposed to break it down into the smallest pieces possible. Have you had your horse's teeth checked lately? Teeth floating is important.

Next, in the stomach, acids and enzymes and endotoxins break the pieces down into a mush. Water is needed for this. Got water?

The next step is tricky. The body has to absorb and break down the mush in the intestines, getting all the vital nutrients it can while moving it along to pass out the back door. There are only so many hours for this to occur.

Ok, now here are the problems. Colic is a blockage or compaction of this material. Colic can be gas caused by the digestion of feed as well. The causes of colic include the following:

1. Teeth in bad shape causing matter too large to be improperly digested.
2. Too much rich food (i.e. hay, grain, lush pastures) when the horse is not used to it.
3. Other medical problems that can bring stress onto horse's overall system.
4. Damage or birth defect in digestion system.
5. Not enough water to drink, or no drinking water.
6. Large infestation of worms.
7. A horse just wormed – dead worms causing a blockage!
8. Massive trauma, kick or buck to stomach area of horse.
9. Most importantly, causes vary and can strike at anytime! Colic is just easy and possible.

Signs of colic range from:

- Lack of energy
- Loss of appetite
- Constant uneasiness
- Biting at the stomach area
- Kicking at the stomach area
- No gastric sounds (gurgling or bubbling)
- Rolling on ground

- Moaning sounds
- Lowered head
- Lack of color in gums
- No recent bowel movements
- Swishing of tail

If you suspect colic, even just minor symptoms, you need to contact your veterinarian ASAP. All horses, no matter what age or sex, even mules, are equally at risk of colic. After contacting your doctor, try to keep the horse from rolling on the ground. Rolling can cause a twisting of the intestines. Keep the horse moving. Walking is good. If the horse wants to lie down, try to prevent this from occurring.

Do not force feed water or any type of oil down the horse's mouth. Wait for the veterinarian. Always have at least two different phone numbers on hand just in case one vet is not available. You'll have a second choice of a vet for help.

If you're concerned about a horse becoming "colicky" contact your local horse doctor for further information on prevention and cure. You and your vet are in much better position than anyone else to decide what's best for your horse's health.

The purpose of this chapter is to give you the basic signs and symptoms, plus what to do in case you suspect colic.

Education is always easier when you personally desire to learn and accept new information. I hope this and the following chapters will do just that... Wet your appetite to go out and learn. Remember, ultimately it is you, the horse's owner, who must make the final decision on all of your horse's needs.

Chapter Forty

"Calculating Hay Consumption"

How much does your horse eat? How many days' supply of hay can your barn hold? How much money will it take to get your horses through the winter? What number of bales will you lose due to leaky roofs? Calculating hay consumption is one of the must popular questions I am asked on my radio show, at a horse clinic and even at the grocery store and gas station. Every horse owner really wants to know how to count hay bales.

Not everyone has the luxury of a nice big barn that can store all the hay you could ever use in a year's time. As hard as it may be to believe, the largest majority of horse owners only have enough storage space for 40 to 75 bales of hay. Forty small square bales of hay does not go far. Consider the fact that most horse owners own more than one horse. There is a big difference between having a "pretty good" idea of how much hay you use to really putting a hard number to that question.

First let me set a base standard to go by. For this chapter we will use a basic small square bale of hay wired tied or twine tied. The national average is 55 to 70 pounds in weight. Hay bales out West consist of three wires and weigh in at around 125 to 150 pounds. Small square bales back East average 45 to 55 pounds. The Midwest is the closest to the average, baling squares that average 55 to 65 pounds. I certainly do not want to leave anyone out, so I will say there is always a neighbor, cousin or someone you know baling hay that weighs about a ton a piece; we're talking about your average small square bale deceivingly weighing as much as four bales!! ☺ I hate trying to carry one of those stupid things!!

What confuses horse owners is that most information that is available on feeding hay and storing hay is based on weight. Also, all nutritional and diet information is based on weight as well. This is why I've chosen this subject for the book. To help put things in a simpler everyday way.

When horse owners ask a nutritionist how much hay to feed their horse, the nutritionist gives them the correct answer. Only the answer and response goes like this:

Question:	How much hay should I feed my horse?
Answer:	Depending on your horse's body weight. A horse generally requires 0.015% roughage everyday. This can be as much as 15 – 18 pounds per day.
Question:	Ok, but how big of a "slab", "square", "flake", "section", "piece", "slice", or my personal favorite hay bale measurement "Look at my hand, bigger or smaller than my hand?"

Answer: Well, I'm not sure. Those types of measurements really cannot be calculated due to bales weighing differently and the various varieties of grass that gets baled.

Final Response: Thank you for your help.

After an exchange like that the horse owner turns to their friend and says "smart guy/ but he doesn't do any chores!!" Let me try to put this into an easier formula for you to calculate hay usage.

First considering I have no idea what age, breed or size of horse you own, not to mention activity level and condition of your horse's teeth, I will pick a middle of the road formula. I'll qualify this further by stating that I'm going to assume for this formula that your horse is a saddle horse 5 to 15 years of age, 15.2 hands tall and weighing in at a petite 1,000 pounds, in other words, your basic riding buddy.

Now for the formula: a horse of this stature will consume a small square bale of hay averaging 55 to 70 pounds in three days. That is a bale being divided into six equal sections. This horse is being fed mornings and nights. So two meals a day gives us three days out of one square bale of hay. Here is how it breaks down using this average formula.

(1 Horse)

1 bale divided by 3 days = 1/3 of a bale a day.
(1/3 divided by 2 meals a day) = 6 meals out of a bale.
365 days in a year divided by 12 months = 30.51 days in each month at an average.

30.51 days divided by 3 days per bale = 10.13 bales needed each month.

That means one horse can be fed

10.13 bales a month multiplied by 12 months = 121.5 bales a year.

So our average horse for this formula eats a little over 10 bales a month. You can then take how many horses you own and multiply from there to get your total year's consumption of small square bales. Don't worry. I'm not even close to finishing this chapter. Now let's talk about the everyday factors that can change the above formula. I'm going to list things you need to take into account when calculating hay usage.

1. Horses that are overweight require less hay. It's called a diet!!
2. Young and older horses do not consume as much hay as a horse in its prime. That is why they have "complete" feeds for younger and older horses.
3. Horses that have access to pastures for grazing require less hay in their daily feed ration. Remember green grass is roughage. It's just not baled yet.
4. Despite the overwhelming majority of rumors, grain does not keep a horse warm in the winter. What keeps a horse warm in the winter is the eating and digesting of roughage. So really cold winters means more hay.
5. Take the opposite approach of what #4 states: warmer winter means less hay needed to keep a horse warm.

6. Activity levels of horses. In other words, performance horses burning lots of calories and energy need more hay than a horse just hanging around the barn.
7. Pregnant mares and mares that are nursing foals. No hay, no baby, or possibly no milk.
8. You always lose a certain amount of hay to broken bales and sloppy chores. Also horses do waste a certain amount of their hay each meal.
9. Bad weather, leaky roofs, poor storage practices.
10. Mice, rats, cats, skunks, possums, chickens, etc... etc... etc. Basically you chalk it up to country living.
11. The next-door neighbor or friend who just wants to borrow a bale here and there.
12. You may buy or sell a horse. This can change your consumption dramatically either way.
13. Any hay left over from the previous year. If it was stored properly, it can be considered like money in the bank.
14. Hay that was baled early and wet that molds and rots away. This can really affect your consumption rate.
15. Someone steals your hay; it happens a lot, easy money or free feed for someone else's horses.
16. Relatives!!!! (Refer to #15)!!! Hey!! Come on. I know some of you are saying this one is your biggest hay consumption problem.

I think I have given you a pretty good foundation for calculating your hay consumption. Now I realize I did not cover round bales, big square bales, loose-stacked hay or cubes. I could go into these topics right now. Yet the biggest number of horses in this country are eating small squares. Besides, believe it or not I'm working on another book. This is the only hint I'll mention.

Chapter Forty-One

"Points of Interest"

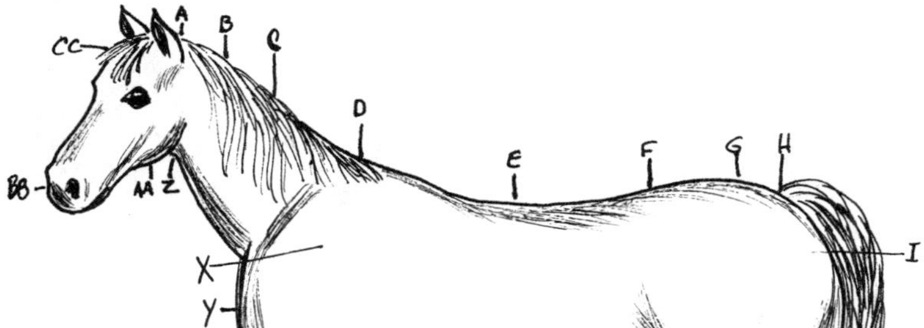
A
B
C
D
E
F
G
H
I
J
K
L
M
N
O
P
Q
R
S
T
U
V
W
X
Y
Z
AA
BB
CC

FULL HORSE

A. Poll
B. Mane
C. Crest
D. Withers
E. Back
F. Loins
G. Croup
H. Dock

I. Hindquarters
J. Thigh
K. Hock
L. Chestnut
M. Fetlock
N. Heel
O. Hoof Wall
P. Pastern

Q. Coronet
R. Cannon
S. Knee
T. Forearm
U. Girth
V. Flank
W. Breast
X. Point of Shoulder

Y. Shoulder
Z. Throat
AA. Jaw
BB. Muzzle
CC. Forelock

This sketch displays a proper fitting halter.

You want there to be 1 7/8" to 2 1/4" of free space between the halter and the horses head.

This applies to the total fit of the halter.

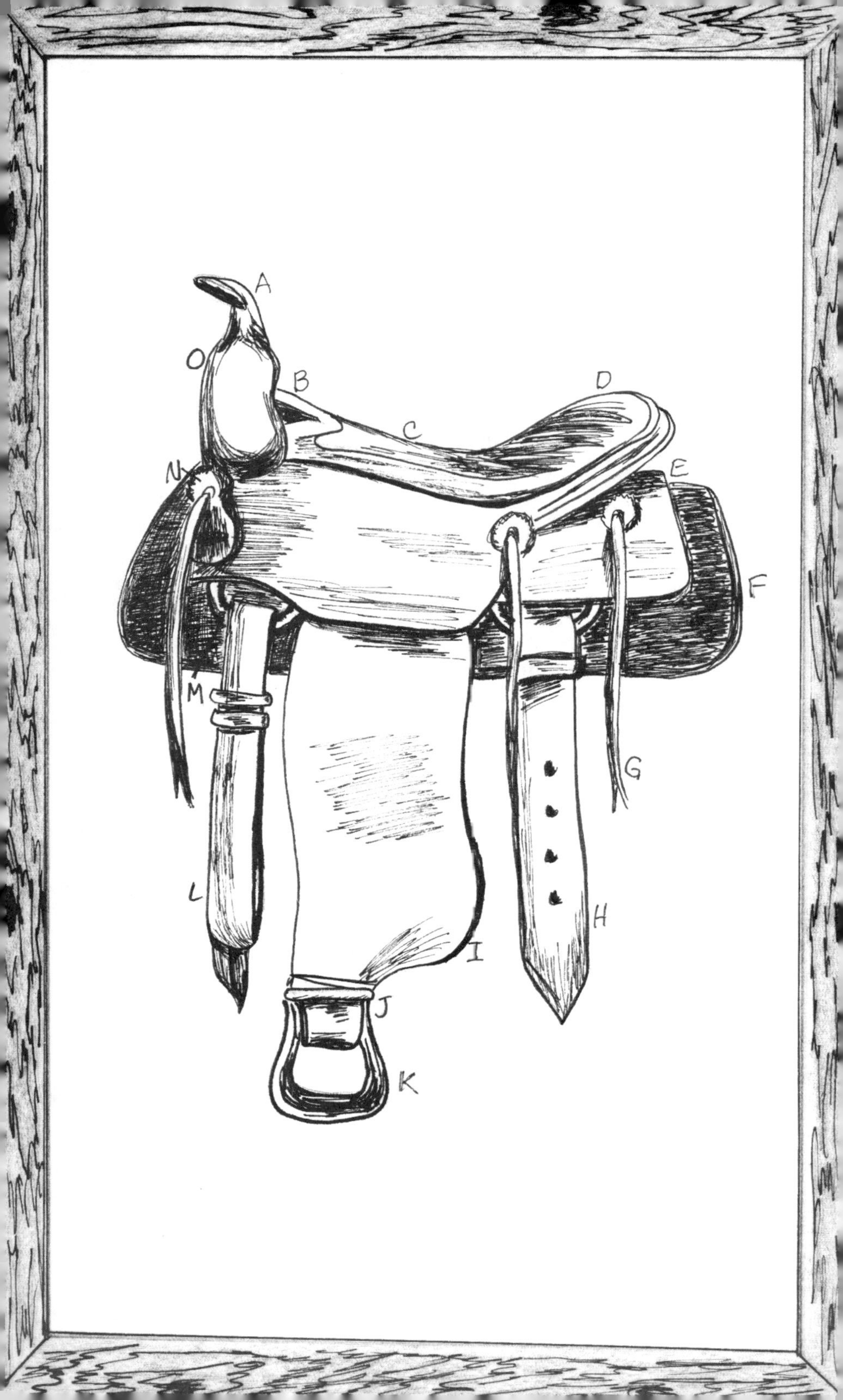

A
O
B
D
C
N
E
F
M
G
L
H
I
J
K

Western Saddle

A. Horn

B. Fork

C. Seat

D. Cantle

E. Upper Flank Skirt

F. Skirt

G. Strings

H. Flank Billet

I. Fender

J. Stirrup Straps Leather

K. Stirrup

L. Latigo

M. D Ring

N. Latigo Carrier

O. Pommel

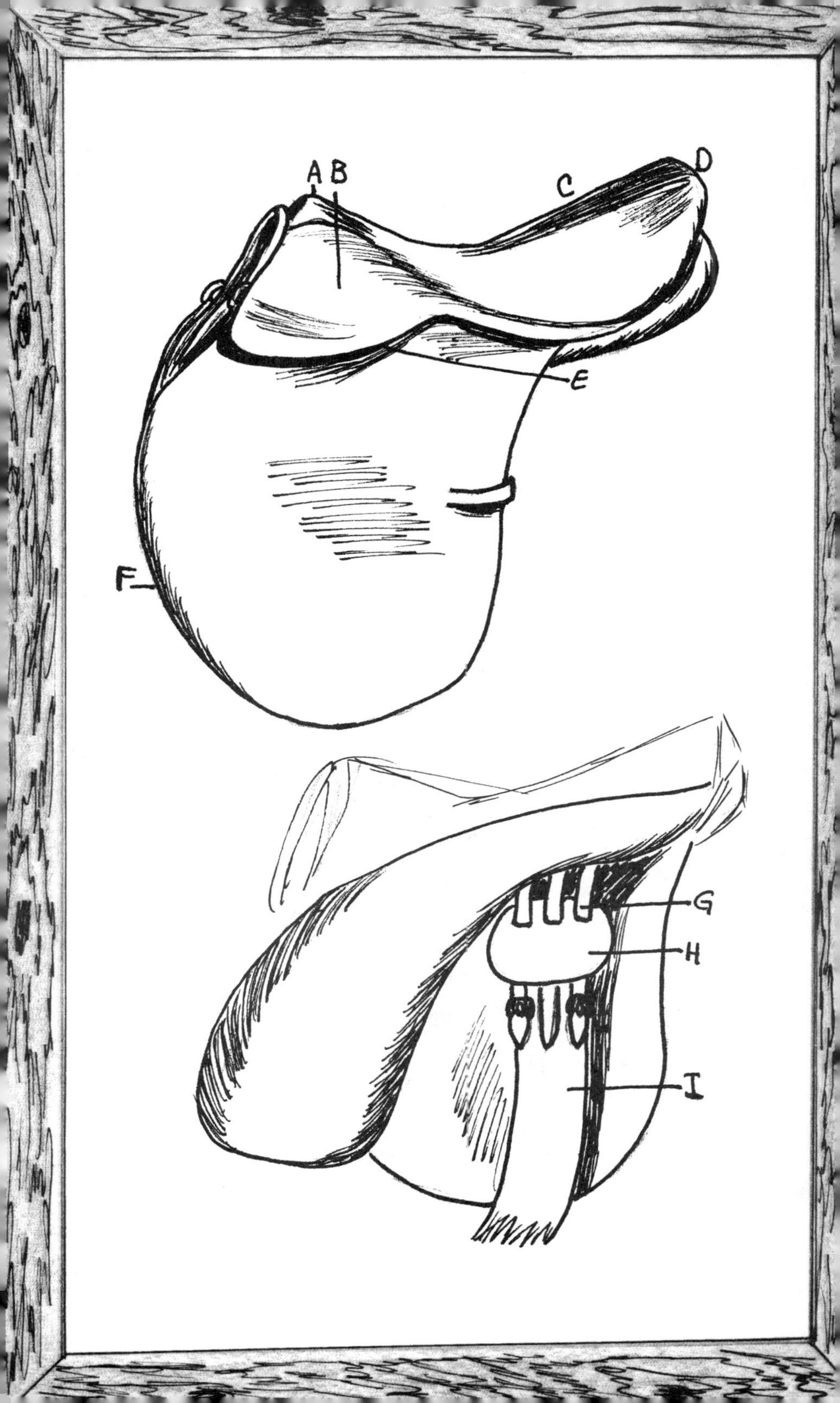

A
B
C
D
E
F
G
H
I

English Saddle

A. Pommel

B. Skirt

C. Seat

D. Cantle

E. Stirrup Bar (Under Skirt)

F. Knee Roll

G. Billets

H. Billet Flap

I. Girth

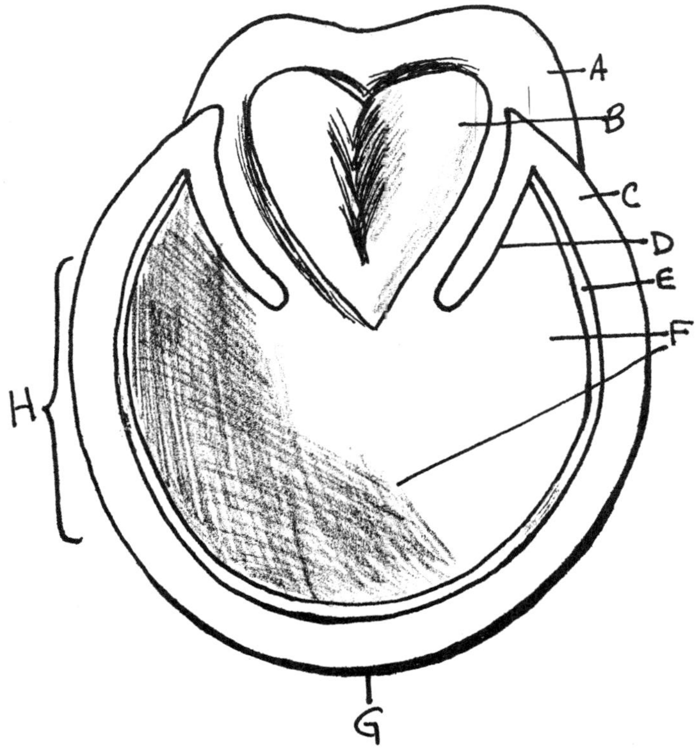
A
B
C
D
E
F
F
H
G

Hoof (Bottom View)

A. Heel

B. Frog

C. Wall

D. Bar

E. White Line

F. Sole

G. Toe

H. Quarter

The diagram on the next page gives you a perspective on how a horse looks and travels with ....

1. Toed out feet
2. Toed in feet
3. Feet that are flat and true

Several things can cause these different looks and foot movements such as trimming inconsistencies, birth defects or choric injuries.

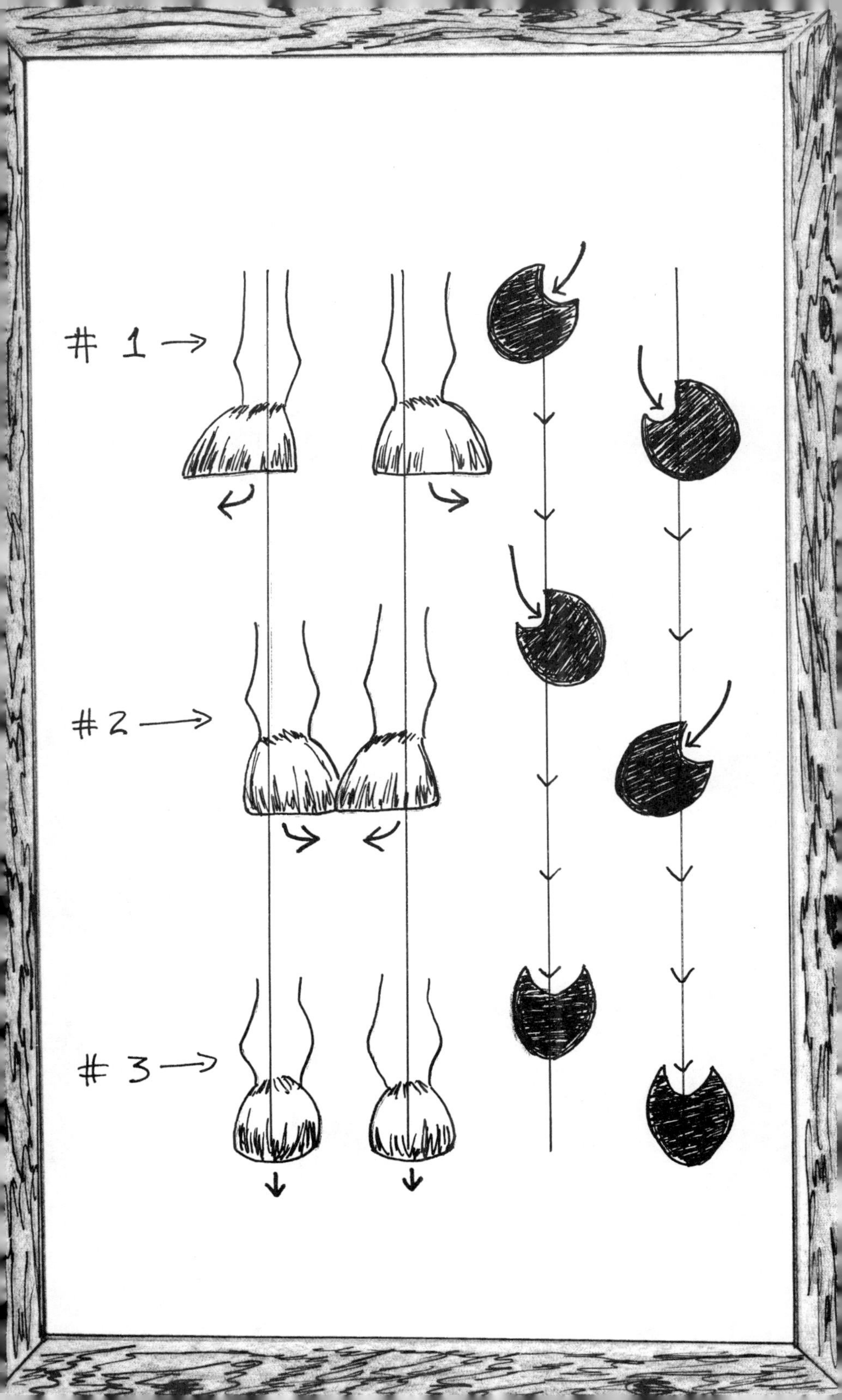

1
2
3

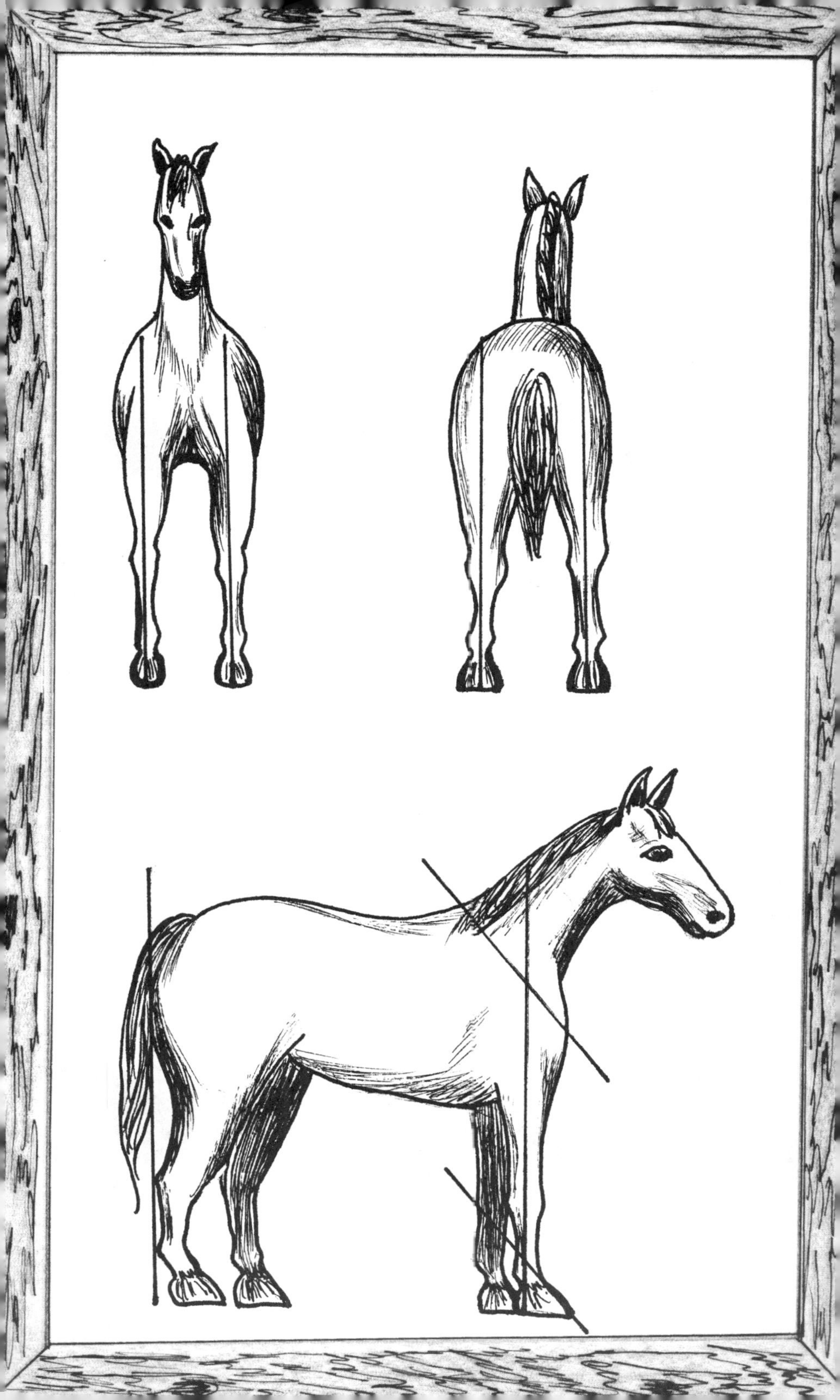

After Trimming

A. Shoulders square and inline with forearms and knees straight down to the fetlock joints through the center of the hoof.

B. The back is square and inline with the stifle, the hocks and the fetlock joints through the center of the hoof.

C. The lines depict proper angles of the shoulder and the rear leg along with the hoof and the pastern angles.

❖ When a horse's feet are trimmed properly and the horse is standing up straight and true you should be able to see these lines of angles on your horse. A wood yardstick or broom handle will help you to gauge your horse's confirmation after the feet have been trimmed.

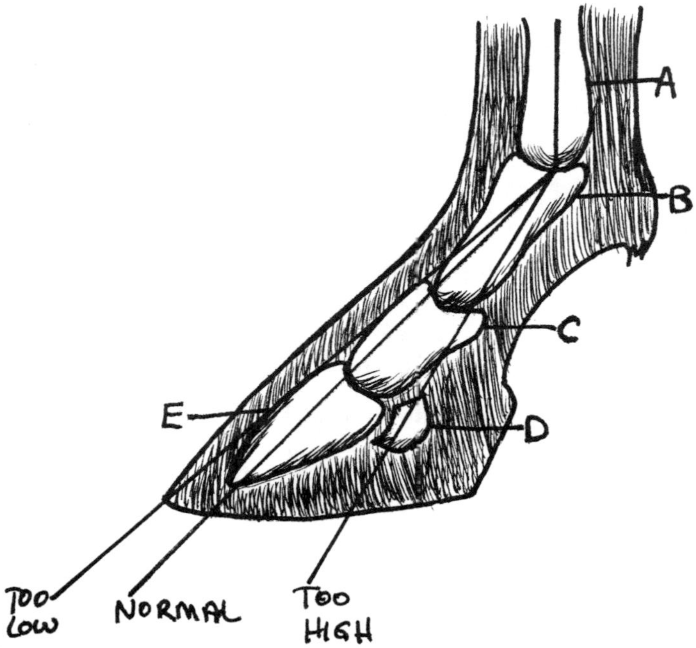
A
B
C
D
E
TOO LOW
NORMAL
TOO HIGH

Hoof With Bones

A. Cannon Bone

B. Pastern Bone

C. Coronary Bone

D. Navicular Bone

E. Pedal Bone

HEIGHT CHART (APPROX.)

Wither Height

40 Inches	=	10 Hands
49	...	12.1
50	...	12.2
51	...	12.3
52	...	13
53	...	13.1
54	...	13.2
55	...	13.3
56	...	14
57	...	14.1
58	...	14.2
59	...	14.3
60	...	15
61	...	15.1
62	...	15.2
63	...	15.3
64	...	16
65	...	16.1
66	...	16.2
67	...	16.3
68	...	17
69	...	17.1
70	...	17.2
71	...	17.3
72	...	18

WEIGHT CHART (APPROX.)

Heart Girth Weight

30 Inches	=	100 Pounds
40		200
45.5		300
50.5		400
55		500
58.5		600
61.5		700
64.5		800
67.5		900
70.5		1000
73		1100
75.5		1200
77.5		1300
80.5		1400
83.5		1500
85.5		1600
87.5		1700
90.5		1800
93.5		1900
95.5		2000
97.5		2100
100		2200
103.5		2300
105.5		2400
107.5		2500

BJ's Final Thought

Normally books do not contain a final thought. If I have had the privilege of your company at one of my horse clinics or we have shared a Sunday evening together on the radio, you will admit: rarely am I conducting myself like other horse professionals. Sometimes others in my chosen profession forget what it's like just owning horses because you love them. It's all well and good to be articulate and official sounding, yet I always try to make sure I never forget what the real world is like. You are the real world of horses.

The toughest part in regards to writing this book was not making sure I had enough information. On the contrary, it was what to put in the book, also what to leave out. I don't think anyone can decide that other than you. You are the one that I talk to on the radio. You're the one who comes to the horse clinics. It is you that sets the goal that I try to achieve. Do I achieve that goal or not? Well, I don't think that is my decision either. That, too, is for you to decide. It is important that I make sure to convey a very important point for you to consider. A point you may want to always keep in mind when trying to learn more about your horses.

Not everything works for all horses. Not all training methods, not all feeds, and certainly not all drugs and medical procedures; horses are as individual as we are. For example, every person requires different sleep patterns, vitamins or no vitamins, lots of food or very little food to maintain a healthy body. Haircuts, dental care, not to mention physical abilities and life expectancies. It is these individualities that make us so varied and yet still so alike.

Not all horses can turn and burn around barrels. There are horses that will never run down a steer and tighten up a rope. Some horses will only cover a quarter mile grazing very slowly. There are thin horses, and there are fat horses. Yet the key word is horses. Also that key word denotes flexibility, creativity, patience and always willingness to concede that there might be a better way.

I've made mistakes over the years; fortunately, I have learned from them. Also the mistakes never were catastrophic or permanent. They could have easily been disastrous. I hope my experiences and education will make your horse experiences less dramatic and more enjoyable. I started this book with giving you the fundamental philosophy of "Horse Time". If you have not figured it out by now, we are right back where we started. That's right. Horse Time.

In closing, please accept my thanks for reading this book. I hope you found it useful and enjoyable. If you would like another "Box Stall Diary", I'm sure I can provide one. This is Box Stall number 13… it can be an enjoyable and informative experience reading your way down to Box Stall number one.

Thanks again, make sure you take care of yourself and also take care of your horses....

"My name is BJ Rickard; I'm in the horse business. This is part of my personal box stall diary..."

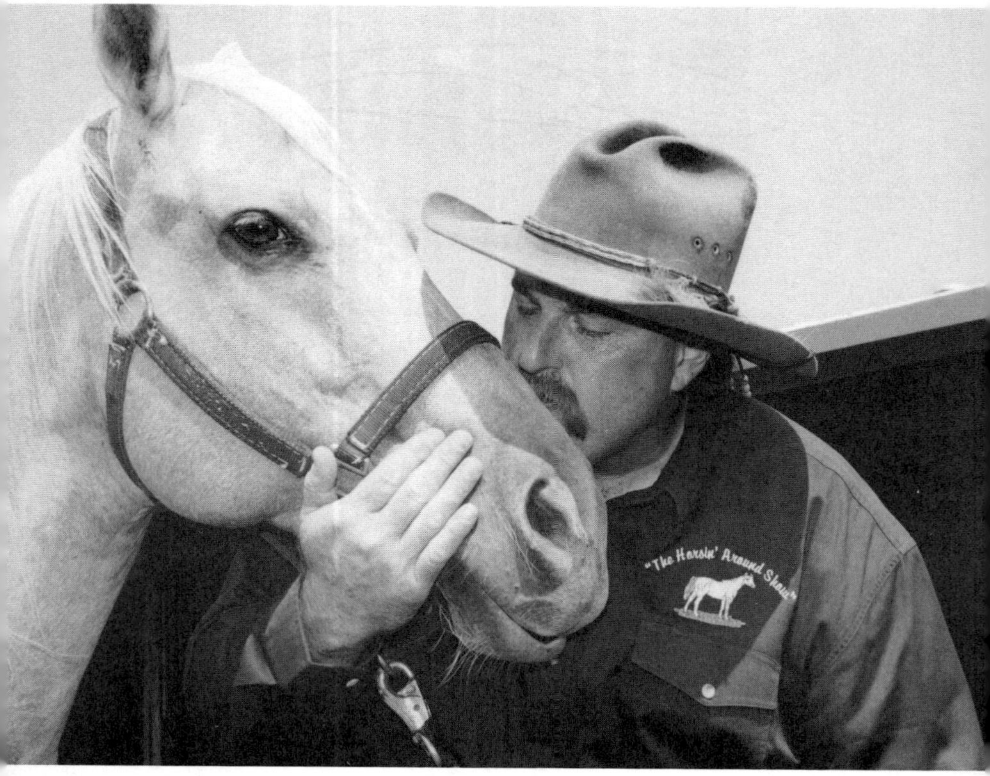
"The Horsin' Around Show"

About the Author

For a number of years, BJ has been involved in many different areas of the horse world. BJ is also recognized as one of the most influential horse professionals in the United States. Horse owners all over the Midwest listen to BJ's radio show, and his informal multi-topic horse clinics are well known and very well respected. BJ has dedicated his life to educating himself on every part of the horse's life, then sharing what he has learned with other horse owners who share his love of horses. When not talking on the radio about horses or standing in front of a crowd sharing his horse experiences, BJ spends every day working with horse owners to help them with their horses' health and training issues. BJ Rickard is one of those rare individuals who can relate to anyone and makes him or her feel right at home. It is this quality along with his well-rounded horse knowledge that puts him in the forefront of the horse business. BJ lives in Oklahoma with his horses, affectionately known to his listeners and fans as the "Lab Rats".

INDEX

A

B

C

D

E

F

G

H

I

J

K

L

M

N

S

T

U

V

W

Y

Z